Design of Reinforced Concrete Buildings to Resist Blast

Niall MacAlevey

Printed by CreateSpace, an Amazon.com company.

Design of Reinforced Concrete Buildings to Resist Blast

To Wen

Design of Reinforced Concrete Buildings to Resist Blast

CONTENTS

1 7
Introduction 7
   1.1   The nature of blast..7
   1.2   Effect of Confinement on explosion 9
   1.3   Free air detonation 11
   1.4   Specification..13
   1.5   Commercial Buildings....................................15
   1.6   Cost of hardening..15
   1.7   How can structures be designed for blast? ..16
   1.8   Static vs. Dynamic Loading............................18
   1.9   Blast Parameters 20
2 24
Local Effects 24
   2.1   Close-in blast...24
   2.2   Contact Explosion..27
3 33
Global Effects 33
   3.1   Advantages of Concrete Structures for blast resistance 33
   3.2   Structural Loading...34
   3.3   Energy of Blast 35
   3.4   Members...36

Design of Reinforced Concrete Buildings to Resist Blast

    3.4.1   Cantilever walls 36
    3.4.2   Column 48
    3.4.3   Slab 53
    3.4.4   Detailing 58
  3.5   Analysis 58
    3.5.1   Single Degree of Freedom (SDOF) Model: Assessment of dynamics of problem 60
  3.6   Summary of Design Procedure 66
    3.6.1   Software Available: 67

4 68

Worked Examples 68

  4.1   RC cantilever wall subject to blast from VBIED. 68

  4.2   RC Level 1 slab subject to blast from VBIED in basement. 79

  4.3   RC column subject to blast from VBIED. 91

5 100

Progressive Collapse 100

  5.1   Review of EC2 104
  5.2   Design against terrorist attack 107
  5.3   Summary: 107

6 109

Façade Design 109

  6.1   Sequence of blast effects-VBIED outside building: 109

Design of Reinforced Concrete Buildings to Resist Blast

- 6.2 Masonry walls 110
- 6.3 Glazing 111
  - 6.3.1 Annealed Glass 111
  - 6.3.2 Toughened Glass 112
  - 6.3.3 Laminated Glass 115
- 6.4 Overall cladding behaviour 117
  - 6.4.1 Design of frame 119
- 7 120
- Derivations 120
  - 7.1 SDOF model: Derivation of K_L, K_M, K_{LM}, K_s ... 120
  - 7.2 Formulae for resistance 123
  - 7.3 Formulae for equivalent elastic stiffness 124
  - 7.4 Solution to equation of motion: free vibration 125
  - 7.5 Dynamic Reactions 126
- 8 128
- Further Examples 128
  - 8.1 128
  - 8.2 129
  - 8.3 130
  - 8.4 132
  - 8.5 134
  - 8.6 137
  - 8.7 138

Design of Reinforced Concrete Buildings to Resist Blast

8.8	140
8.9	143
8.10	143
8.11	144
8.12	148

Design of Reinforced Concrete Buildings to Resist Blast

1

INTRODUCTION

1.1 THE NATURE OF BLAST

A blast is the result of a very fast chemical reaction. The explosive often contains the elements Carbon, Hydrogen, Oxygen and Nitrogen, e.g. TNT (trinitrotoluene) is $C_6H_2(NO_2)_3CH_3$.

As the reaction proceeds large quantities of gas (roughly $1m^3$ for each kg of explosive) are generated (Smith and Heatherington 1994). When high explosive is detonated, it causes a shock wave which travels at supersonic speed. The air in front of these gases is expelled. The pressure of the air suddenly increases. This is the "positive phase" of the explosion.

After several milliseconds, the pressure of the gases drops, usually below atmospheric pressure. The air-flow reverses direction. This is the "negative phase". The loading (impulse) from this negative phase (known as 'rebound') may be 50% or more of the positive phase (known as 'inbound').

When a blast wave is unobstructed the pressure is at its "incident" value. Once it strikes a surface in its path the pressure on the surface rises. This process is described as

Design of Reinforced Concrete Buildings to Resist Blast

"reflection". The blast waves engulfs any object in its path.

Consider air molecules represented by spheres with momentum Mv (see Fig. 1.1). The collision with an object is elastic, i.e. the velocity is exactly reversed by the collision. In this case the object has decelerated the 'sphere' from v to zero, and then accelerated it back to v in the reverse direction. Thus the force (change in momentum/time) experienced by the object in the path of the sphere is $2Mv/t$.

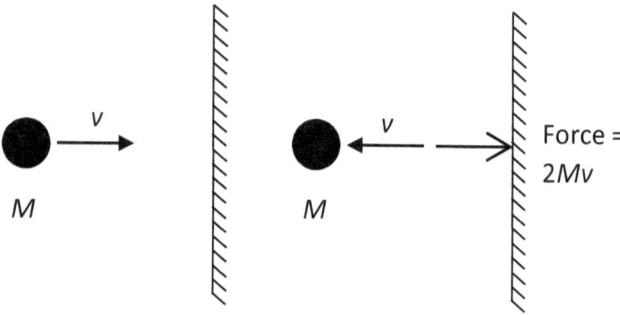

Fig. 1.1: Reflection of spheres

In reality the air compresses (see Fig. 1.2) so the magnification is frequently more than the '2' above.

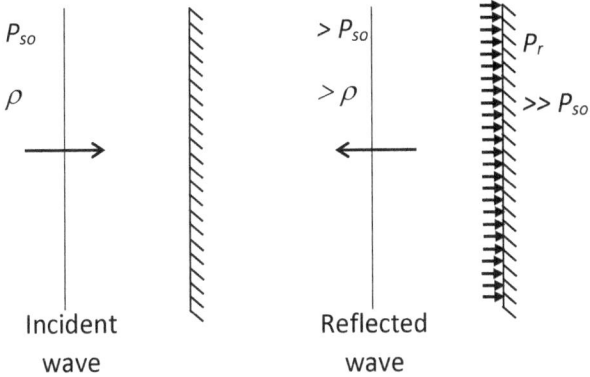

Fig. 1.2: Reflection of blast wave.

When a bomb explodes on or near the ground outside a building, the side of the building facing the bomb is exposed to reflected pressure, the opposite face is exposed to incident pressure, and the roof is exposed to incident pressure. Thus the pressures attempt to crush the building (see Fig. 1.3).

1.2 Effect of Confinement on Explosion

Generally the effect of confinement is to make the effects of the explosion worse. Multiple reflections of the blast wave take place as the wave bounces from surface

Design of Reinforced Concrete Buildings to Resist Blast

to surface. In addition, explosion products (i.e. gas) build-up, further increasing the pressure. (See Fig. 1.4).

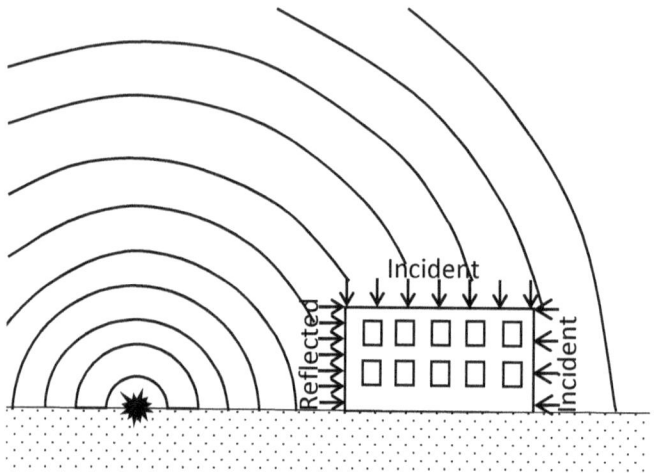

Fig. 1.3: Building engulfed by blast wave

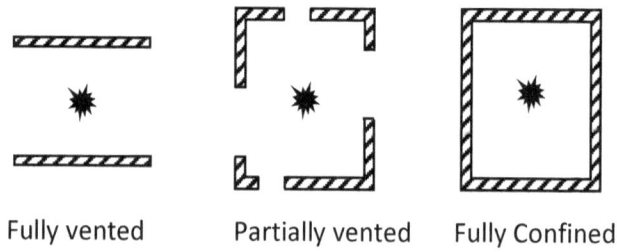

Fully vented   Partially vented   Fully Confined

Fig. 1.4: Degrees of confinement

1.3 Free Air Detonation

A typical detonation of high explosives in a fully-vented environment ("free air") results in loading of the form illustrated in Fig. 1.5.

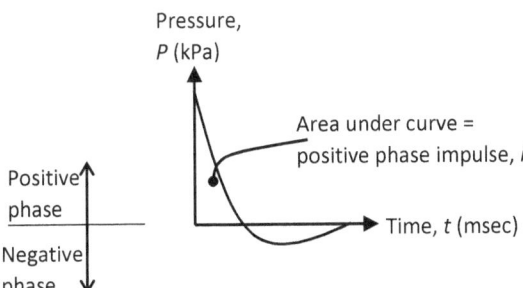

Fig. 1.5: Blast parameters

For example, 100 kg TNT at a distance of 4 m, causes a positive reflected pressure of 12,000 kPa for 0.84 milliseconds (reflected impulse ≈ 5,100 kPa.msec). A millisecond is 1/1,000 of a second. For comparison, the blink-of-an-eye takes about 120 milliseconds. The area of the positive impulse triangle = 0.5*12,000/0.84 = 5040 kPa.msec. Strictly speaking we are concerned here with *specific* impulse ($i = I/t$), i.e. Pressure.time. not simply impulse (I) i.e. Force.time.

Design of Reinforced Concrete Buildings to Resist Blast

Because the blast waves radiate out from the source like concentric spheres, so the effects of the explosion diminish very rapidly with distance (for instance, pressure is inversely proportional to the cube if the radius of the sphere. Hence the expression:

"The cheapest protective hardening measure is STAND-OFF"

(STAND-OFF is distance from the explosion, i.e., if you can ensure the explosion happens further away from the structure, then the hardening measures will be much less costly.) Stand-off is ensured by bollards, planters, etc.

E.g., US Department of Defense rating:

Standard vehicle: 15,000 lbs (6,810 kg)

"K4" => travelling 30 mph (50 kph)

"K8" => travelling 40 mph (65 kph)

"K12" => travelling 50 mph (80 kph)

Specification of barriers includes penetration: "L1" (1 metre), "L2" (1-6 metre), "L3" (6-15 metre).

To illustrate the importance of stand-off: Fig. 1.6 shows the plot of impulse versus stand-off for a charge of 250 kg TNT.

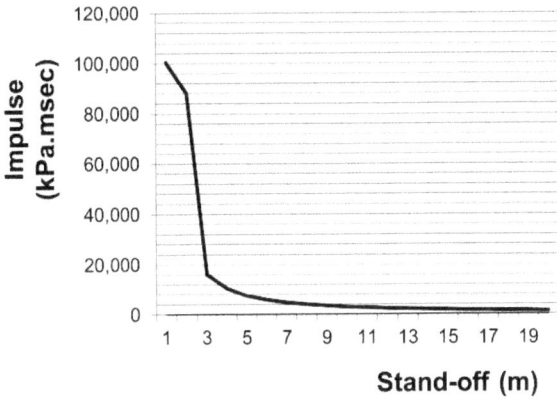

Fig. 1.6: Plot of impulse vs. stand-off for 250 kg TNT.

1.4 Specification

As an example of terrorist threat specification, the Ministry of Home Affairs, Singapore, gives this list of security threats to be considered for most new government buildings (offices, hospitals, museums, electrical substations, etc.) and large private buildings (e.g., hotels, shopping centres), see Table 1.1.

Design of Reinforced Concrete Buildings to Resist Blast

Table 1.1: MHA threat specification

Threats	Methods	Specifics
Vehicle Borne Improvised Explosive Device (VBIED)	Possible vehicle	Cars, vans/pick-ups, lorries, trucks
	Possible weight	Approximately 500 kg TNT_{EQ}
	Possible methods of Delivery	Ramming detonation attack, Timer activated detonation
Person Borne Improvised Explosive Device (PBIED)	Possible weight	Approximately 20 kg TNT_{EQ}
	Possible concealment	Suitcase, haversack, parcel, suicide vest, etc.
	Possible methods of Delivery	Delivered, thrown, motorbike, etc.

Note: Charge is given as TNT equivalent. The explosive may be TNT but this is not so common for terrorist bombs. Single blast event.

The density of TNT is 1,600 kg/m^3 so 20 kg is about a cube of side 230 mm (9"), and 250 kg is a cube of side 540 mm (1.8').

1.5 COMMERCIAL BUILDINGS

Many structures incorporate a basement car park. If there is no control over entry of vehicles, the building is exposed to a VBIED threat from beneath the building itself. This is thus an expensive decision from a security/blast point of view.

The alternative is to avoid car parks near the structure: people park off-site and are bussed to the building.

1.6 COST OF HARDENING

As mentioned above, to reduce the cost of hardening, where possible stand-off is maximized. However this invariably means the cost of land and perimeter protection increases. Thus the solution is a compromise.

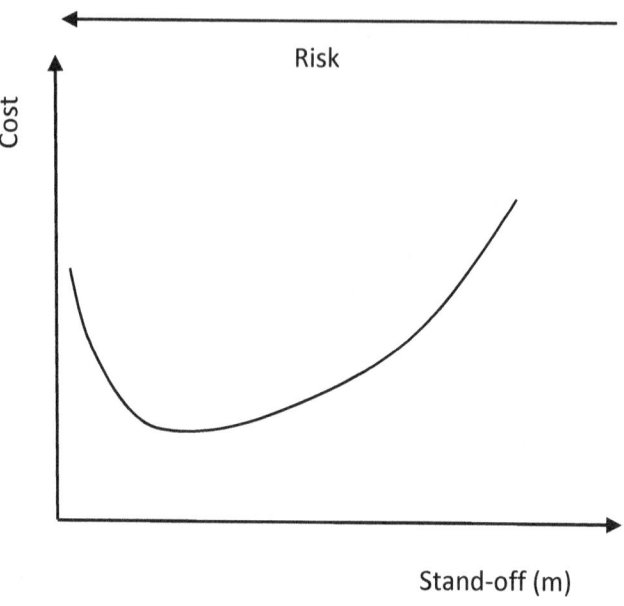

Fig. 1.7: Cost of building vs Stand-off

1.7 How can structures be designed for blast?

Although the pressure from a blast wave is often large in magnitude, structures can often be designed to resist their effects. Pressures of thousands of kPa can be resisted **because they are of very short duration!** The fact that the load is suddenly applied and therefore dynamic, actually helps us.

Design of Reinforced Concrete Buildings to Resist Blast

Very short duration compared to what?

For high explosive detonation, the time duration of the blast tends to be short in comparison to the **natural period of vibration** of the element of structure that experiences the blast (see Fig. 1.8).

This sort of loading is described as **impulsive** (imparting a momentum to the structure) and maximum deflections will occur after the net load has diminished to zero.

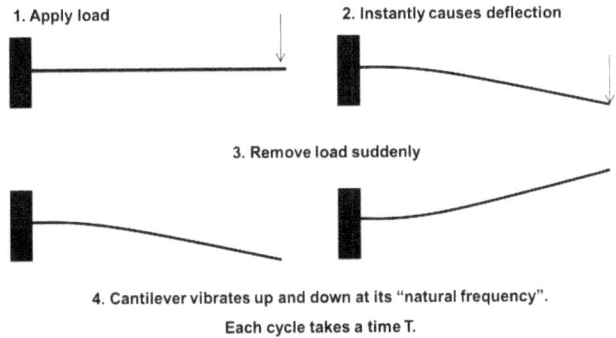

Fig. 1.8: Natural frequency (n) and period (T)

We are perhaps all familiar with dynamic loading. Consider this example: suddenly get on a weighing scales. The needle will vibrate for a few seconds and then come to rest at your static mass. The highest number it reaches as soon as you suddenly apply your weight is theoretically about twice your static mass.

This suddenly applied loading is said to have a Dynamic Load Factor (DLF) of about 2, i.e. 2 times your static weight.

1.8 Static vs. Dynamic Loading

The following rule of thumb is helpful:

$$\frac{\text{Load Duration, } t_d}{\text{Natural Period, } T} > 5 \Rightarrow \text{the load is STATIC}$$

otherwise it is DYNAMIC

Typically for a blast t_d is around 1 millisecond while the natural period of the structural member might be 10 milliseconds: thus $t_d/T \approx 0.1$. The Dynamic Load Factor for this sudden load is **less than 1**.

The influence of this ratio for triangular loading is show by Fig. 1.9 below (taken from Biggs 1964).

Design of Reinforced Concrete Buildings to Resist Blast

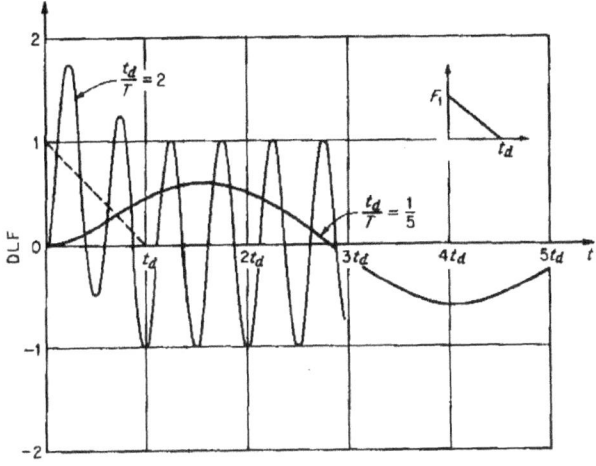

Fig. 1.9: Effect of t_d/T ratio on DLF

For conventional gravity loadings the structure is loaded very gradually (i.e. load duration, t_d is very large). T is typically less than 1 second, so t_d/T very high. Thus the effects are STATIC. If the load is DYNAMIC then the effects are = STATIC x DLF. As mentioned already, in a high explosive blast DLF is typically much less than 1.

We can summarize the loading regimes as follows:

Impulsive	Dynamic	Static
$t_d \ll T$	$t_d \approx T$	$t_d \gg T$
DLF $\ll$ 1	DLF 1-2	DLF = 1

1.9 BLAST PARAMETERS

The values of the blast parameters (pressure, impulse etc.) can be estimated from so-called "spaghetti charts" such as the following for TNT. See Fig 1.10. These diagrams are a record of empirical data obtained from many blast tests conducted in the open air. (DoD 2008).

Note: m = explosive mass (kg); R = stand-off (distance from explosion) (m).

P_r = reflected pressure

P_{so} = incident pressure

I_r = reflected impulse

I_s = incident impulse

t_a = wave arrival time

t_o = duration of positive phase

U_s = speed of wave

Design of Reinforced Concrete Buildings to Resist Blast

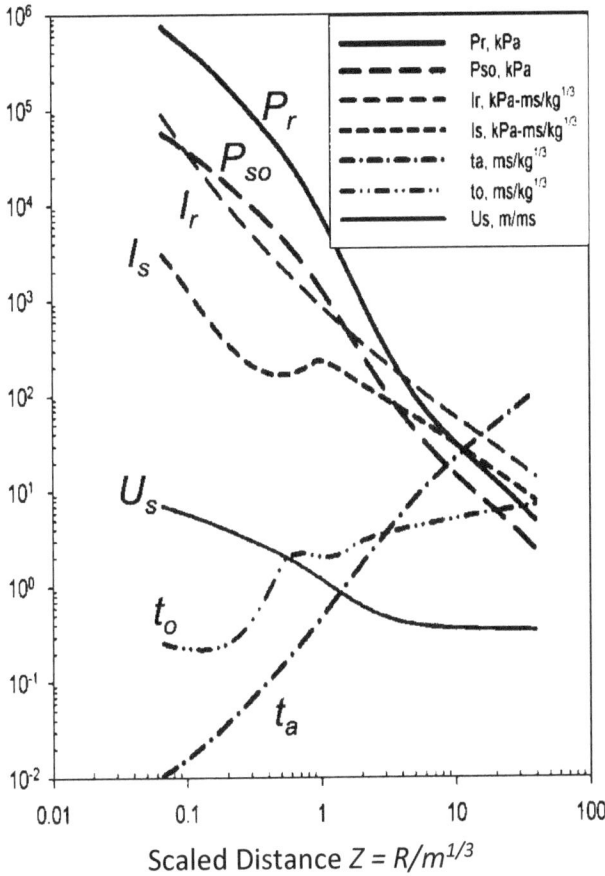

Fig. 1.10: Open-air TNT blast

Design of Reinforced Concrete Buildings to Resist Blast

Example:

Consider 100 kg TNT at a stand-off of 10 m. Find value of positive phase incident and reflected pressures and impulses.

Solution: $R = 10$ m, $m = 100$ kg

=> Scaled distance, $Z = R/m^{1/3} = 10/100^{1/3} = 10/4.63 = 2.15 \text{ m/kg}^{1/3}$

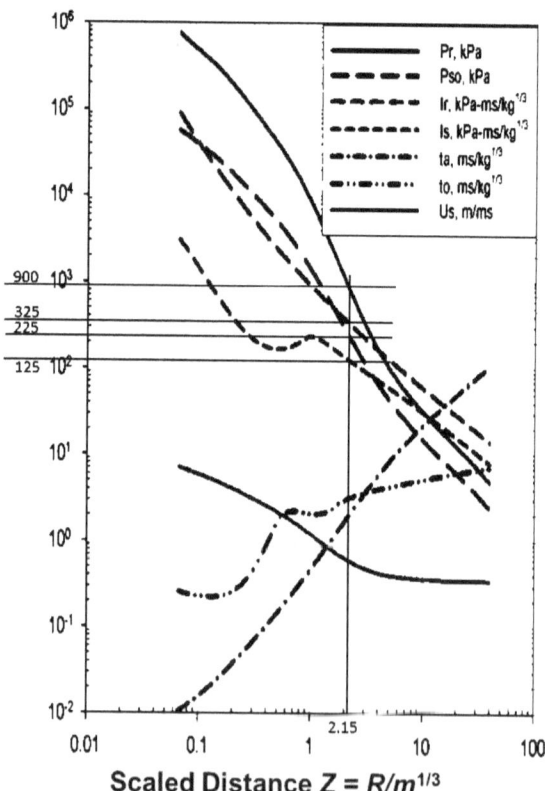

Design of Reinforced Concrete Buildings to Resist Blast

Reading from the graph:

- Incident pressure, P_{so} = 225 kPa;
- Incident impulse, I_s = $125 \times m^{1/3}$ = $125 \times 100^{1/3}$ = 580 kPa.msec
- Reflected pressure, P_r = 900 kPa;
- Reflected impulse, I_r = $325 \times 100^{1/3}$ = 1,508 kPa.msec.

Note: This chart gives the inbound (i.e. positive phase) blast impulse; there is a rebound (i.e. negative phase) impulse too. This rebound impulse is always less than the inbound impulse, and is typically about 50% of it. Thus there is a significant reversal of loading. The impulse is related to the total energy of the blast, and it is this that the structure should be designed to resist.

Available software: Blast Parameters (i.e. *P, I, t* etc):

1. "3D Blast": purchase from Applied Research Associates (http://www.ara.com/products/software.htm)
2. "ConWep": distributed free to qualifying parties (must be American citizens).

2

LOCAL EFFECTS

2.1 CLOSE-IN BLAST

If the blast occurs in close proximity to the structural members, it can cause failure by spalling or breaching. This material failure is the subject of this chapter.

When a blast wave impacts a concrete member, a compression wave passes through the concrete. When this compression wave reaches the far surface of the concrete, it is reflected as a tensile wave.

If the tensile strength of the material is exceeded then the rear face will experience spalling (Fig. 2.1). If large enough this can completely breach the structural element. The process of destruction by spalling is known as **brisance**.

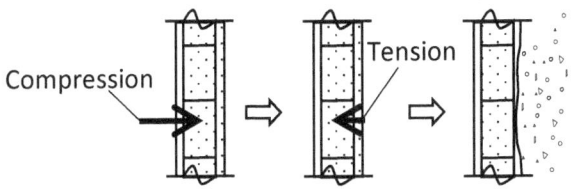

Fig. 2.1: Spalling of concrete

Blast tests allow the local effects to be assessed. For example, Fig. 2.2 shows the diagram by Hader:

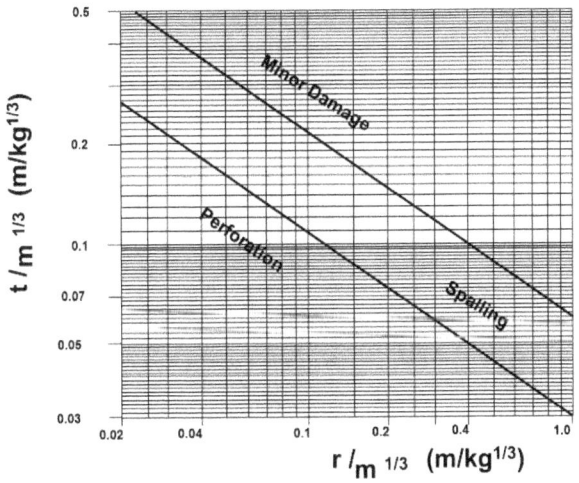

Fig. 2.2: Local Effects of a Blast

Design of Reinforced Concrete Buildings to Resist Blast

Where t is the wall thickness, r is the stand-off and m is the mass of the charge. Apparently, the quantity of reinforcement has little effect when the explosion is close-in. Krauthammer (2008) seems to agree.

Example:

Using the Hader diagram, assess the local effects of a 100 kg TNT charge 3 m from a 200 mm thick wall.

Solution:

We must quantify the following parameters:

- $t/m^{1/3}$, where t is the thickness of the concrete and m is the mass of the charge.
- $r/m^{1/3}$, where r is the stand-off.

$t/m^{1/3} = 0.2/(100)^{1/3} = 0.2/4.63 = 0.04$

$r/m^{1/3} = 3.0/(100)^{1/3} = 3.0/4.63 = 0.65$

Now we see where these two points lie on the diagram.

Design of Reinforced Concrete Buildings to Resist Blast

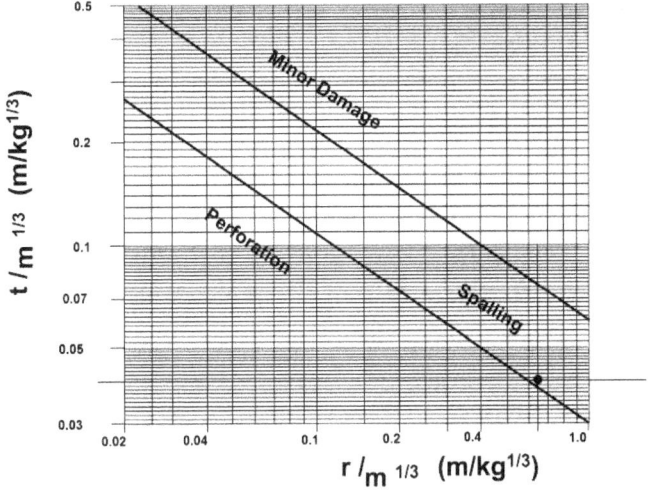

The diagram tells us that there is likely to be severe spalling and as the point is close to the perforation line, some risk of perforation.

If the area behind the wall is used for mass occupation of people (more than $0.1/m^2$) the wall is certainly unacceptable. If, however, the area is a plant room with non-critical equipment, or a little used corridor then it possibly is satisfactory.

2.2 CONTACT EXPLOSION

Up to now we have been considering the effect of an explosion caused by a bomb likely carried by a vehicle (vehicle borne improvised explosive device, VBIED)

Design of Reinforced Concrete Buildings to Resist Blast

where stand-off is metres or so. The minimum stand-off for a VBIED is about 1.5 m.

Now we will consider the effect of a smaller charge that is hand-carried (a person borne improvised explosive device, PBIED) and placed in contact with an RC member.

The PBIED considered for design may be anything from 2 kg (suicide vest)-20 kg (suitcase) TNT.

The Hader diagram gives us some information, but suppose we want to know more about approximately how much damage will be caused by a blast very near the surface of the member.

The contact blast generates so much pressure and high temperature that the concrete is locally destroyed (e.g. Fig. 2.3).

Design of Reinforced Concrete Buildings to Resist Blast

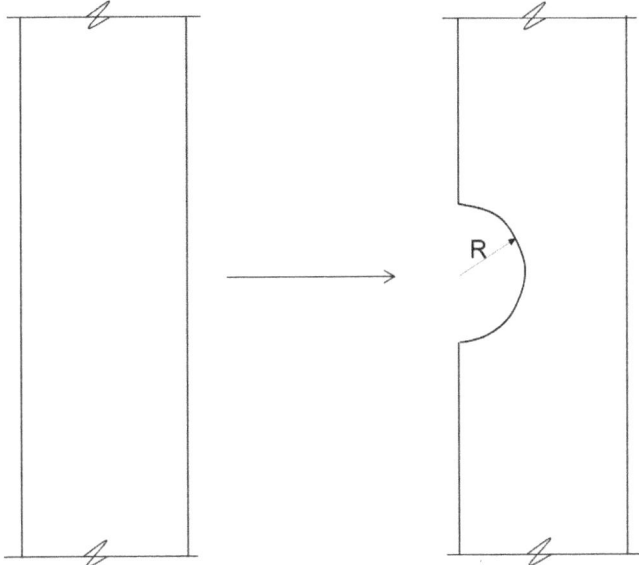

Fig. 2.3: Damage caused by contact explosion.

Usually, empirical data from full-scale blast tests is used to assess these effects.

The Israeli Corps of Engineers–Manual for Design of Protective Structures, 1987, and the US Field Manual 5-250, 1992, provide detailed information to assess the portion of the member that is likely to be removed by the blast.

A sphere of explosive is imagined to be touching the concrete member. See Fig. 2.4.

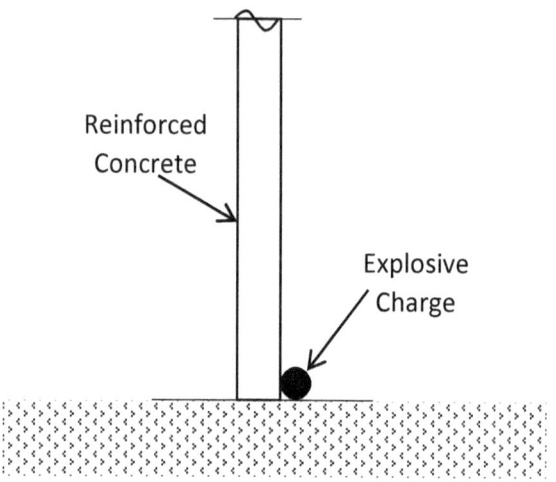

Fig. 2.4: Contact Explosion

The empirical formula for estimating the radius of breaching of a reinforced concrete wall is:

$$R = \sqrt[3]{(m/16KC)}$$

where:

R – radius of breaching of the concrete wall (meters)

m – the charge mass (kilograms)

K – material coefficient as shown in Table 2.1.

C – confinement coefficient as shown in Table 2.2.

Design of Reinforced Concrete Buildings to Resist Blast

Table 2.1: Material Coefficient, K

Thickness of Concrete Wall (m)	K
0.3 or less	1.76
0.45 – 0.75	0.96
0.9 – 1.35	0.80
1.5 – 1.95	0.63
2.1 and above	0.57

Table 2.2: Confinement Coefficient, C

C	Charge Location	Description
1.0		Deep Underground Explosion
2.0		Explosion on surface with partial confinement
1.8		Above ground explosion with no confinement
3.6		Explosion on surface without confinement

Design of Reinforced Concrete Buildings to Resist Blast

Example:

Consider a reinforced concrete column of 800 mm x 800 mm cross section, exposed to a contact IED of 10 kg TNT. Find the value of the breaching radius R. The charge is placed on the ground adjacent to the column.

Solution:

$m = 10$ kg; $K = 0.9$; $C = 3.6$

$R = \sqrt[3]{m/16KC}$

$R = \sqrt[3]{10/16 \times 0.9 \times 3.6} = 0.58$ m

Thus 0.58 m of the section will be removed. This leaves 0.8-0.58 = 0.22 m of intact column. Many of the reinforcing bars will survive, but their restraint against bucking may be lost. Depending on the axial load on the column, this may or may not be enough to mean the column can take at least some load until it is repaired.

3

GLOBAL EFFECTS

This chapter examines the effect of the blast when the structure deflects and attempts to absorb the energy of the blast.

3.1 ADVANTAGES OF CONCRETE STRUCTURES FOR BLAST RESISTANCE

Blast load durations are very short compared to other extreme structural loads, and are often shorter than the natural period of many structural elements. Because of this blast waves do not excite the mass of the structure in the same way as longer duration, cyclic loading (e.g. from an earthquake).

Hence we can often analyze a structural element as if it were isolated from the structure.

The mass of concrete structures provides a direct resistance to the blast wave. Their weight is roughly twice that of an equivalent steel building. (Concrete building density typically 400 kg/m^3 compared to 200 kg/m^3 for steel). They are heavy so they have large inertia. Thus they are more difficult to set in motion.

Design of Reinforced Concrete Buildings to Resist Blast

The inherent continuity of RC structures allows properly detailed structures to develop alternate load paths and resistance mechanisms.

Well detailed concrete structures have substantial ductility and energy dissipation characteristics.

Lastly, the potential for fires from explosions is high, both from the explosion itself as well as from secondary fires caused by damage. The inherent fire resistance of concrete is an advantage, particularly due to the potential of having a compromised fire suppression system following a blast event.

3.2 STRUCTURAL LOADING

For conventional loadings (gravity and wind) the structure is usually assumed to remain elastic at service and designed for Ultimate Limit State using a factored elastic bending moment diagram, perhaps allowing for a small amount of plasticity by assuming some moment redistribution. (The 30% limit to redistribution given by codes is to ensure the beam is satisfactory at the Serviceability Limit State (i.e. working load level)).

Now consider the imposition of a very unusual load: blast loading. Survivability is the issue. Clearly there are no Serviceability Limit State requirements (e.g., deflection and cracking). If a structure is exposed to blast loading the pressure imposed on the structure can be (very briefly) thousands of times higher than that due to wind (which is typically 1 kPa). Thus to economically and practically resist this load, full advantage of "plastic design" (also known as "limit design") must be taken. Thus we assess the limit to the capacity of the structure (i.e., the maximum possible capacity).

3.3 ENERGY OF BLAST

Close-in blasts are usually **impulsive**, meaning they impart momentum (Impulse = mass x velocity). The structure should deflect in response to this load and be able to decelerate safely. As the load is not sustained infinitely, there is no need to design for the value of the pressure P of the blast, only the impulse, I.

Now conservation of energy considerations mean:

Energy in = Energy absorbed.

Thus Blast Energy = Kinetic Energy of Structure (structure gets an initial velocity) = Strain Energy of Structure (plastic hinges form due to yielding of the steel and then rotate).

Thus the structure receives an initial velocity = Impulse of blast/Mass of element. (We will assume the loading is purely impulsive for the moment.)

i.e. $v = I/M$

Thus structure has kinetic energy = $½ Mv^2$

Design of Reinforced Concrete Buildings to Resist Blast

Deflection of structure should be sufficient to absorb this energy.

(The structure must bend to absorb the energy. Think of a long reed bending in the wind. It does this so it can absorb the energy. Work is Fd so if d is large then F can be small.)

We will consider the various structural members and examine their requirements to ensure blast resistance.

- Cantilever Walls
- Columns
- Slabs
- Beams
- Shear walls (effects on bracing system).

3.4 MEMBERS

3.4.1 Cantilever walls

Consider the blast wave as represented by a uniformly distributed loading. Then the collapse mechanism is as follows (Fig. 3.1):

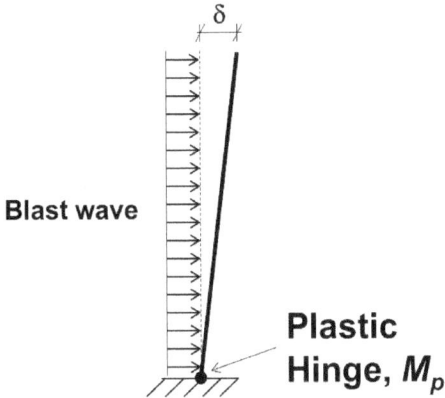

Fig. 3.1: Cantilever wall under blast

The wall bends in response to the blast wave. A cantilever structure is statically determinate so a single hinge is enough to make the structure into a collapse mechanism. Ordinarily this would mean collapse of the structure because ordinarily the load is sustained, but for blast-resistant structures the load has probably already disappeared by the time the hinge appears, so there is no 'failure', just permanent deformation.

The amount of movement required is that needed to absorb the energy of the blast. The plastic hinge is formed at the base of the wall. Note that the plastic hinge is not like a door hinge: more like a rusty hinge, there is resistance to movement (the resistance is M_P). The hinges rotate as the wall deflects. This rotation absorbs the energy.

Design of Reinforced Concrete Buildings to Resist Blast

The actual value of the UDL resisted by the wall is *w*; this UDL is limited by the moment at the plastic hinge, M_P as the UDL must be in equilibrium with M_P. Thus, if the structure is ductile, this is the maximum load the structure can 'see'.

We can now design the structure to resist **this** UDL (i.e. that implied by the size of M_P). Thus the load to be designed for *w*, is determined only by M_P rather than directly by the applied blast pressure.

Of course we cannot just arbitrarily choose any value of M_p. We must limit the permanent deformation after the blast 'event' is over, i.e. we must have M_p large enough to absorb the energy of the blast.

3.4.1.1 *Flexure*

At the hinge, yielding of the steel is allowed, but crushing of the concrete is not (since this represents ultimate failure of the section). The concrete must remain intact during the rotation so that compression due to flexure can be supported. Thus the lower part of the wall must be ductile: the steel must yield first, allowing the hinge to rotate. This means it is essential that the neutral axis, *x* is small, otherwise the concrete will crush. Fig. 3.2.

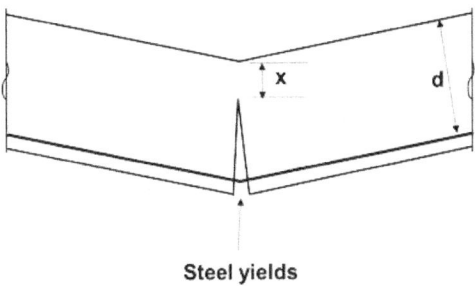

Fig. 3.2: Formation of Plastic Hinge

Recall that ensuring that the value of x is small means that at the ULS the steel yields before the concrete maximum strain is reached (i.e., before crushing of the concrete). See Fig 3.3.

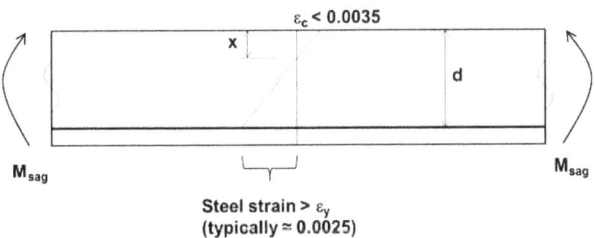

Fig. 3.3: Ultimate value of neutral axis depth x.

EC2 gives a guideline for limiting the value of x, it says that a frame element can be designed plastically if $x/d \leq 0.25$. No further checks are necessary. (Reinforcement must be type B or C that is medium or high ductility).

Design of Reinforced Concrete Buildings to Resist Blast

The wall section is usually **symmetrically** reinforced for several reasons:

- In order to ensure that the value of x is low;
- Rebound is of the same order as the inbound;
- Reduce the chance of site errors.

Load factors are usually taken as 1.0 as:

- The loading is unusual;
- Materials are loaded quickly (the strain rate is high) and are stronger as a result (given less time to respond properly). See Fig 3.4.

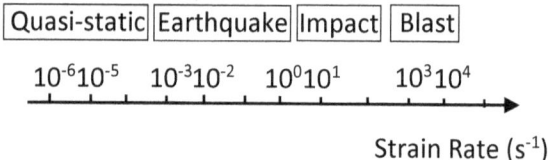

Fig. 3.4: Strain rates of loading types

E.g. 'Quasi-static' Loading: e.g., compression test of concrete takes about say 2 minutes (120 seconds): total strain of concrete about is 0.0035. Thus strain rate = $0.0035/120 = 3 \times 10^{-5}$ (s^{-1})

At the high strain rates characteristic of blast loading (100-1000/sec) experiments show:

- Concrete compressive strength may increase by a factor of up to 3 and tensile strength by up to around 4.
- Steel yield strength can be doubled, and ultimate tensile strength (i.e. breaking strength) increased by 50%.

However, the factors recommended for design are more conservative. These are as follows (Table 3.1):

Table 3.1: Design Stresses

Stress	Concrete	Steel
Bending	1.25	1.20
Compression	1.15	1.10
Shear	1.00	1.10

Thus the factors for design in flexure to EC2 are as shown in Fig. 3.5.

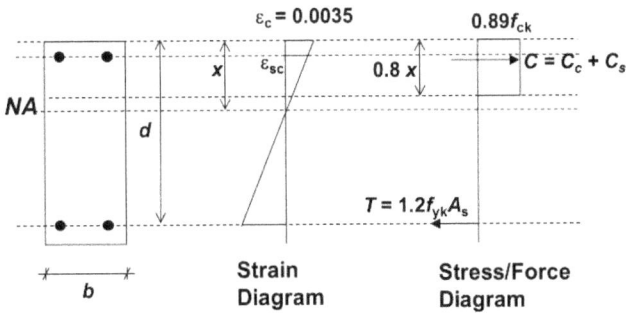

Fig. 3.4: Factors for Flexural Design to EC2

Design of Reinforced Concrete Buildings to Resist Blast

Example

Calculate an approximate value of the flexural capacity of a 3 m high cantilever blast wall, the section of which is shown below. The material strengths to be used are as follows: f_{ck} = 45 N/mm² and f_{yk} = 417 N/mm² (ε_{uk} > 5%). Find the value of w, the static UDL. Cover 30 mm.

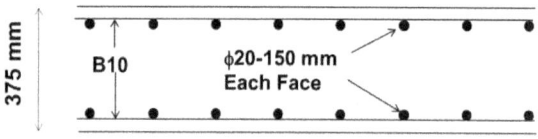

Solution

Section has equal tension and compression reinforcement. Thus an approximation for the moment capacity can be obtained by ignoring the concrete in compression. The lever arm will be the distance between the centres of the two layers of reinforcement i.e. z = 375-2(30)-2(10)-20 = 275 mm.

Consider a 1000 mm length of wall.

Tensile force, T = (1.2x417)(314)(1000/150) = 1,046 kN

Moment capacity, M_p = Tz = 1046(0.275) = 288 kNm.

This is a conservative approximation.

Notice that providing equal tension and compression reinforcement means the value of x is likely to be low. Note that links must be added in order to restrain the compression bars as well as for 'shear'.

Design of Reinforced Concrete Buildings to Resist Blast

Next we will determine the value of the UDL w. As ultimate is approached, reinforcement yields allowing a hinge to form at the support. The moment capacity of this hinge is M_p.

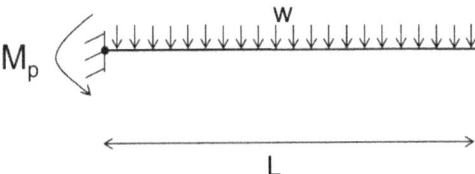

From statics $wL^2/2 = M_p$

Thus $w = 2M_p/L^2 = 2(288)/3^2 = 64$ kN/m

3.4.1.2 Diagonal Shear

Diagonal shear is a brittle failure and so must be avoided. Thus we must ensure <u>no diagonal shear failure</u> occurs, as shear failure is brittle. Just as in flexure, the value of shear expected is a function, not of the applied blast load, but of the value of the moment of resistance of the column hinge (as the latter decides the value of the UDL w). If the moment of resistance, M_p is large, then so is the shear demand.

Design of Reinforced Concrete Buildings to Resist Blast

3.4.1.3 Direct Shear

Blast tests have revealed that there is another brittle failure mechanism we must guard against: failure at the face of the support due to <u>direct shear</u>. The wall is at risk of literally sliding off the supports (Fig. 3.5).

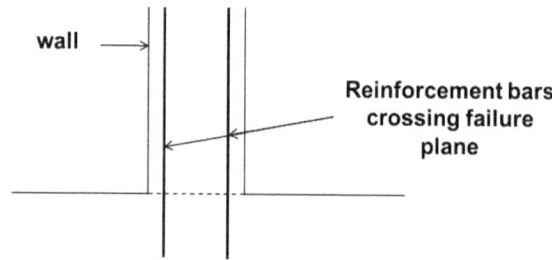

Fig. 3.5: Direct Shear Failure at Face of Support

Reinforcement bars resist this sliding. The resistance mechanism is known as 'shear friction' (Fig 3.6). The failure plane is rough so any lateral movement involves vertical too. Thus *vertical* reinforcement, A_h is used to resist *horizontal* force, F (the direct shear).

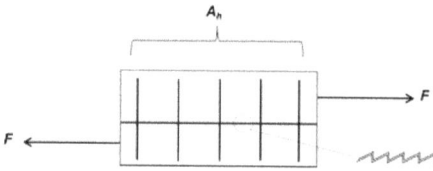

Fig. 3.6: Shear Friction Mechanism

Especially if there is no axial load on the wall, it is sometimes necessary to use diagonal bars to resist the blast load (Fig. 3.7).

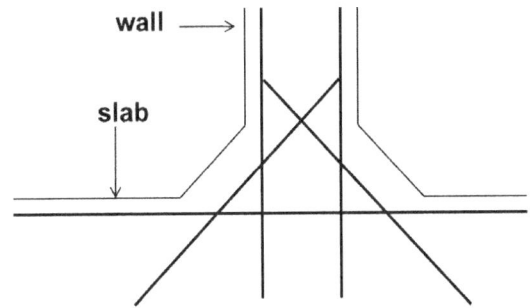

Fig. 3.7: Diagonal Reinforcement

3.4.1.4 Limit to support rotation

Normally the support rotation for a cantilever wall is limited to about 2-4 degrees. This limit refers to the amount of permanent deformation necessary to allow the hinge to absorb the energy of the blast. The limit chosen (a bigger limit meaning there is less protection but the solution is more economical) reflects the amount of damage which can be tolerated. The hinge should be able to rotate by this amount without failure of the concrete. (Fig. 3.8).

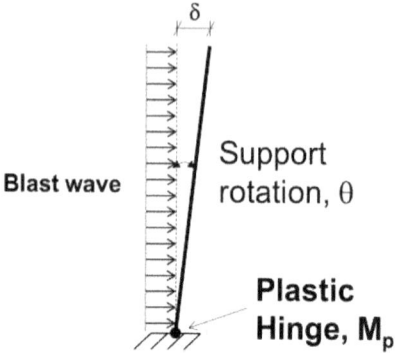

Fig. 3.8: Support Rotation = Index of Damage

3.4.1.5 Protection Category

Large rotations (more than about 2 degrees) effectively assume the cover concrete can disintegrate (Fig. 3.9). The section is then known as a "Type 2" section, and the Protection Category is considered to be level 2 (lower level of protection). The conventional section, where the compression concrete remains intact, is a "Type 1" section. The Protection Category if the section is Type 1 is considered to be level 1. (Fig. 3.10)

Design of Reinforced Concrete Buildings to Resist Blast

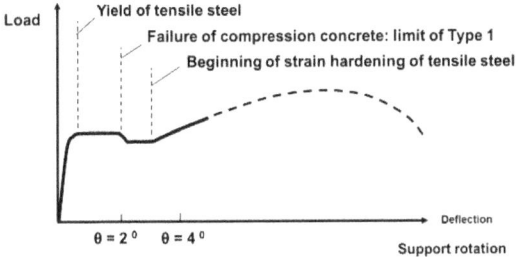

Fig. 3.9: Flexural Behaviour

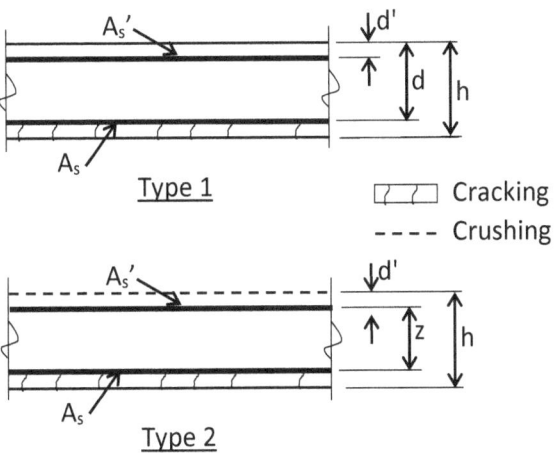

Fig. 3.10: Section Types

Design of Reinforced Concrete Buildings to Resist Blast

3.4.2   Column

The big difference between the column and the wall relates to the likely presence of an axial compression in the column. The axial load considered usually is DL + 50% LL (EC2). Note that ACI/SEI 59-11 recommends inclusion of the axial load if the axial stress > $0.1f_{ck}$.

Consider again the pressure wave represented as a UDL. The column bends in response to the blast wave. As always, the amount of movement required is that needed to absorb the energy of the blast.

A structure with fixed ends requires three hinges in a line in order to fail (i.e., a mechanism). Three plastic hinges are formed in a column (upper, near mid-height and lower). The hinges rotate as the column deflects. See Fig. 3.11.

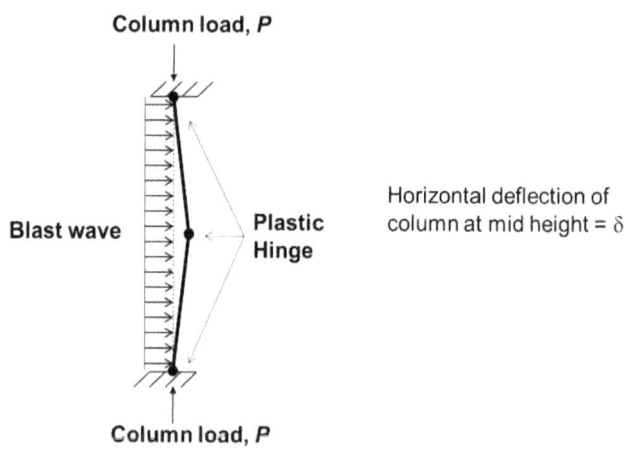

Fig. 3.11: Column under blast load

Design of Reinforced Concrete Buildings to Resist Blast

3.4.2.1 Flexure

At the hinge, yielding of the steel is allowed, but crushing of the concrete is not. The sections where the plastic hinges are likely to form must be ductile: the steel must yield first allowing the hinge to rotate. Concrete must remain intact so that compression due to flexure can be supported. This means it is essential that the neutral axis, *x* is small, otherwise the concrete will crush.

As before, calculation of the flexural capacity allows for the larger dynamic material strengths. A column section is usually symmetrically reinforced; this means that the value of *x* is likely to be lower and so helps to ensure ductility. The axial column load means there is an axial compression on the section which should be included in our calculations (its effect is to increase *x*). Again, all load factors are usually taken as 1.0. Material factors according to EC2 are those shown in Fig. 3.4.

Example

Given that the moment capacity of a column section (M_p), calculated using load factors of 1.0, dynamic strengths of materials, and an axial load of DL+0.5LL, is 1,000 kNm, find the corresponding value of the static UDL consistent with the hinges of the mechanism. Column clear height between restraints is 5 m.

Design of Reinforced Concrete Buildings to Resist Blast

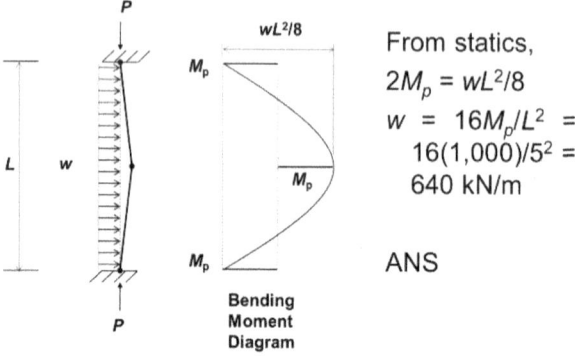

From statics,
$2M_p = wL^2/8$
$w = 16M_p/L^2 = 16(1,000)/5^2 = 640$ kN/m

ANS

3.4.2.2 Diagonal Shear

In addition to ensuring x is small, we must also ensure <u>no diagonal shear failure</u> occurs, as shear failure is brittle. As before, the value of shear expected is a function, not of the applied load (blast), but of the value of w consistent with the values of M_p. Thus we check based on the UDL w. These values of M_p are in turn made larger by the presence of the axial compression P.

Higher strength concrete frequently used in columns (e.g., above f_{ck} = 45 N/mm^2) is likely to have a higher tensile strength so is diagonal shear is less likely to be a problem. Nevertheless it is common to have closely spaced links to act as shear reinforcement in the column.

3.4.2.3 P-delta Effect

We must also ensure that the column does not move "excessively". A large value of the column deflection δ results in a large value of extra moment $P\text{-}\delta$. (Fig. 3.12). This would be an extra destabilizing moment for the column, which if uncontrolled could lead to the collapse of the column. Frequently only a deflection equivalent to a $P\text{-}\delta$ moment of 10%-15% of M_p is allowed but judgment is required.

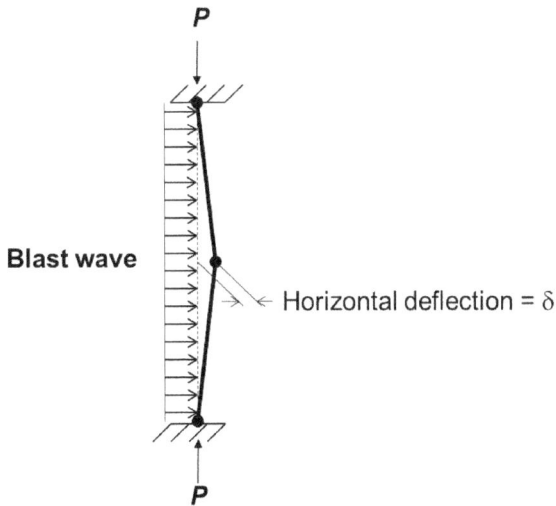

Fig. 3.12: $P\delta$ additional moment

3.4.2.4 Direct Shear

Blast tests have revealed that there is another failure mechanism we must guard against: failure at the face of the supports due to <u>direct shear</u>. The column is at risk of literally sliding off the supports. See Fig. 3.13 and 3.14.

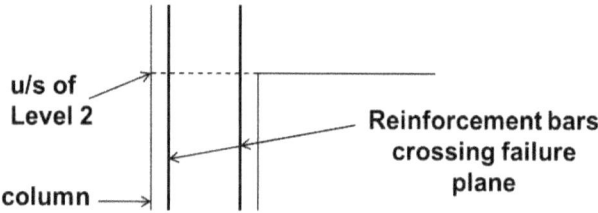

Fig. 3.13: Direct shear failure plane

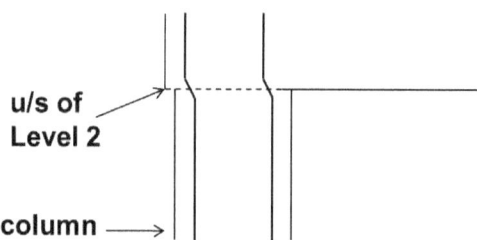

Fig. 3.14: Direct shear failure

The presence of vertical bars and axial compression is beneficial ('shear friction'). However, as <u>cold-joints</u> are present at both ends of the column we need to be especially careful of this connection.

Design of Reinforced Concrete Buildings to Resist Blast

3.4.2.5 Detailing

Laps in the area likely to be exposed to the blast should be avoided. Similarly, welding should be avoided.

3.4.2.6 Confined Concrete

Advantage may be taken of the presence of links which act to confine the core concrete. The increase in the concrete strength means the section is more ductile (lower x) and the moment capacity (M_p) is increased. A method is described by Paulay and Priestly (1992). See section 3.2.2 of that book.

3.4.3 Slab

A slab is a _flexural_ member (usually developing only bending moments and shear forces to resist the load; rarely is there significant axial load.)

Consider again the pressure wave represented as a UDL. Usually the blast originates about 1m above ground level. Thus the blast applies an _upward_ UDL to the slab above. The slab bends upwards in response to the blast wave by an amount needed to absorb the energy of the blast. A structure with fixed ends requires three hinges in a line in order to fail (i.e., a mechanism).

Design of Reinforced Concrete Buildings to Resist Blast

Three plastic hinges are formed in a slab (left support face, right support face and near mid-span). The hinges rotate as the slab deflects. Small deflection => Elastic BMD applies (Fig. 3.15); larger deflection => Plastic BMD applies (only difference is the fixed values of moment, M_p, at the plastic hinges). See Fig. 3.16.

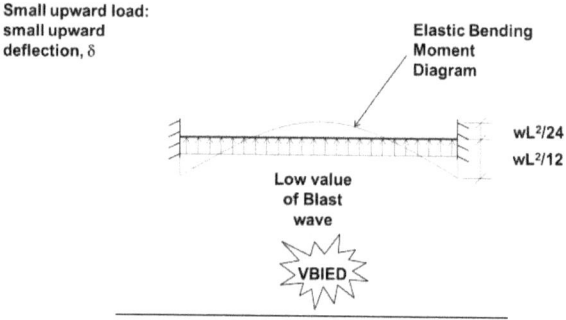

Fig. 3.15: Elastic Bending Moment Diagram

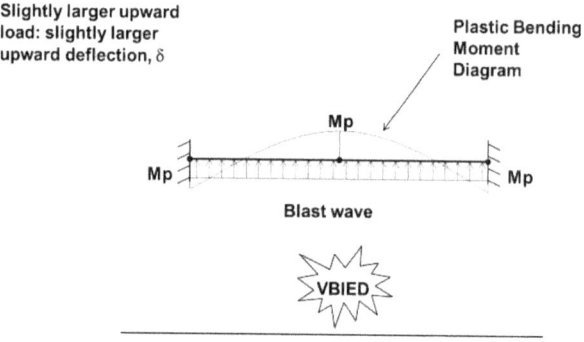

Fig. 3.16: Plastic Bending Moment Diagram

Design of Reinforced Concrete Buildings to Resist Blast

3.4.3.1 *Flexure*

As before, at each hinge, yielding of the steel is allowed, but crushing of the concrete is not, i.e. the concrete must remain intact so that compression due to flexure can be supported. Thus at the hinge locations the RC must be ductile: the steel must yield first, allowing the hinge to rotate. This means it is essential that the neutral axis, x is small, otherwise the concrete will crush.

Continuous top reinforcement is necessary. As well as being required for blast, during the service life of the structure, this helps with the control of deflections and shrinkage cracks due to restraint. The presence of substantial compression reinforcement also means the value of x is likely to be low. As before, load factors are taken as 1.0. Materials are at their dynamic strengths and so the material factors are as Fig. 3.4.

3.4.3.2 *Diagonal Shear*

In addition to ensuring x is small, we must also ensure no diagonal shear failure occurs, as shear failure is brittle. If the moment of resistance is large then so is the shear demand. Hence it is not unusual to have links in the slab. As well as providing shear reinforcement, links restrain the compression reinforcement from buckling.

Design of Reinforced Concrete Buildings to Resist Blast

3.4.3.3 Direct Shear

Blast tests have revealed that there is another failure mechanism we must guard against: failure at the face of the support due to <u>direct shear</u>. The slab is at risk of literally sliding off its supports (Fig. 3.17 and 3.18).

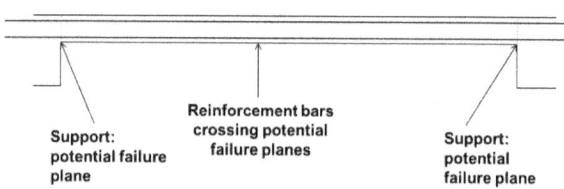

Fig. 3.17: Slab (links not shown)

Reinforcement bars resist this sliding ('shear friction'). Thus the connection should be checked to ensure sufficient reinforcement crosses it.

Fig. 3.18: Direct Shear Failure

3.4.3.4 Limit to Support Rotation

Normally the support rotation is limited to about 2 degrees but if there is sufficient lateral restraint to ensure tension membrane action then up to 8 degrees is allowed (tension membrane action, i.e., cable action, can be allowed if the reinforcement is continuous and the support can resist the horizontal reaction). This limit chosen affects the amount of damage which can be tolerated (fig. 3.19).

Blast load upward load: large upward deflection, δ
If δ is more than half the slab thickness then Compression Membrane (arch) action likely; If more than the slab thickness then Tension Membrane (catenary) action likely.
Can add more than 50% to carrying capacity.

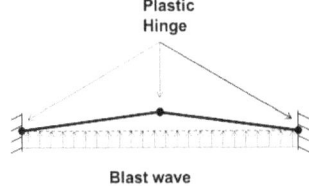

Fig. 3.19: Response limits

3.4.3.5 Beam

Comments as for slabs apply to beams too. Diagonal shear is more likely to be a problem. Links should be of closed type.

Design of Reinforced Concrete Buildings to Resist Blast

3.4.3.6   Shear Walls

It will be recalled that the function of the shear walls is to resist the lateral loads on the building. The lateral loads are transferred to the shear walls by the floors acting as diaphragms.

Blast imposes a large lateral load on buildings. However as the wave impinges on the building, some of the mass of the building is engaged in resisting the blast. These inertia forces mean that there may actually be little effect on the shear walls.

3.4.4   Detailing

Symmetrical reinforcement preferred since rebound due to negative phase and in order to help minimize site errors.

Elements exposed to direct effects of blast usually receive links. The main purpose of these links is to restrain the compression bars and to confine the concrete so greater strains can be achieved.

3.5   ANALYSIS

Problems in structural dynamics typically involve significant uncertainties, particularly with regard to

Design of Reinforced Concrete Buildings to Resist Blast

loading characteristics. Thus complex methods of analysis are often not justified. It is a waste of time to employ methods having precision much greater than that of the input of the analysis.

"Static" shear is the term given to the set of shear forces in equilibrium with the resistance UDL W. "Dynamic" shears are the actual shears experienced during the dynamic loading (see "derivations"). The following table gives dynamic shears.

We usually use the "static" shear for design as it tends to be larger than the dynamic shear (at least for impulsive load).

Table 3.2: Dynamic Shears

Edge Conditions and Loading Diagrams	Dynamic Reaction		
	Elastic	Elasto-plastic	Plastic
$F = pL$	0.39W+0.11F	-	0.38W+0.12F
	0.69W+0.31F		0.75W+0.25F
	0.36W+0.14F	0.39W+0.11F	0.38W+0.12F
	Left: 0.39W+0.11F	0.39W+0.11F	0.38W+0.12F

Design of Reinforced Concrete Buildings to Resist Blast

3.5.1 Single Degree of Freedom (SDOF) Model: Assessment of dynamics of problem

Each structural member is replaced by a simplified model, and examined one member at a time. The Single Degree of Freedom (SDOF) model is normally used (as this is the simplest dynamic model). An "equivalent load", w_{equ} is applied to an "equivalent mass", M_{equ}, which is connected to a spring of "equivalent stiffness", k_{equ}. Only one direction of movement in response to the load is allowed, i.e., x. See Fig. 3.20.

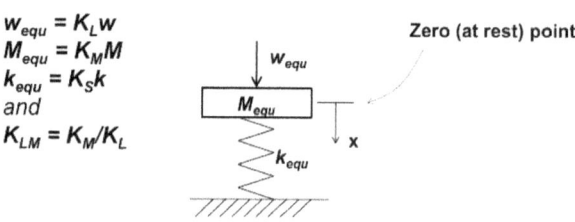

Fig. 3.20: SDOF model

The SDOF model is a simplification:

- Each portion of the actual continuous structural element has multiple degrees of freedom and its own equation of motion;
- We replace this continuous structure by a single point mass. We are interested only in the response of this single mass.
- The actual load applied is not a point load.

Design of Reinforced Concrete Buildings to Resist Blast

3.5.1.1 Energy Balance

An impulse imparts initial velocity I/M to the structure:

The Kinetic energy (KE) thus imparted to system is

$$\tfrac{1}{2} Mv^2 = \tfrac{1}{2} M(iA/M)^2 = \tfrac{1}{2} i^2 A^2/M \quad \ldots\ldots\ldots\ldots(1)$$

since $v = I/M$

where M is the mass, v is the velocity and i (= I/A) is known as the specific impulse where A is the area of the member exposed to the blast.

To bring the structure to rest the KE must be absorbed as strain energy (SE).

Energy = Force x deflection.

Thus the larger the value of deflection that can be obtained, the less structural resistance that needs to be provided. (Also it can be seen that, since $v = I/M$, more mass M is beneficial as less energy needs to be absorbed.) See Fig. 3.21.

Design of Reinforced Concrete Buildings to Resist Blast

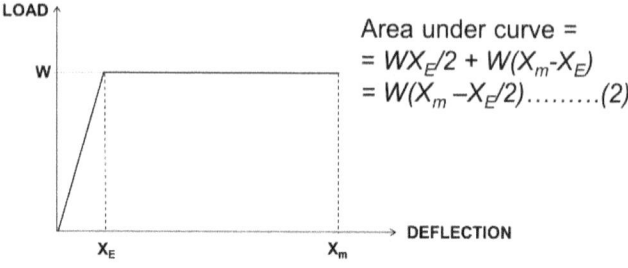

Fig. 3.21: Strain Energy

Thus equate "external" and "internal" energy (1) and (2):

$$½ \, i^2 A^2 / M = W(X_m - X_E/2)$$

This is known as the **Impulse equation.**

As we use a SDOF system for design, it is further modified (see derivations) to be:

$$\boxed{½ \, i^2 A^2 / MK_{LM} = W(X_m - X_F/2)}$$

The following table shows values of K_{LM}. The design of any structure can be carried out once this equation is solved. See Table 3.3.

Table 3.3: Conversion factors

Edge Conditions and Loading Diagrams	K_{LM}		
	Elastic	Elasto-plastic	Plastic
(simply supported, uniform load)	0.78	-	0.66
(cantilever, uniform load)	0.65	-	0.66
(fixed-fixed, uniform load)	0.77	0.78	0.66
(fixed-simple, uniform load)	0.78	0.78	0.66

3.5.1.2 Summary

$$\tfrac{1}{2} i^2 A^2 / M = W(X_m - X_E/2)$$

where

i is the specific impulse (= I/A)

A is the area of the structural element (exposed)

Design of Reinforced Concrete Buildings to Resist Blast

M is the mass of the structural element

W is the resistance (load consistent with hinges)

X_m is the maximum displacement

X_E is the displacement at yield of the steel

Table 3.4 shows values of w for various cases.

Table 3.4: Values of Resistance (w)

Edge Conditions and Loading Diagrams	Ultimate Resistance, w
(simply supported, uniform load w)	$8M_{SAG}/L^2$
(cantilever, uniform load)	$2M_{HOG}/L^2$
(fixed-fixed, uniform load)	$8(M_{HOG} + M_{SAG})/L^2$
(propped cantilever, uniform load)	$4(M_{HOG} + 2M_{SAG})/L^2$

Design of Reinforced Concrete Buildings to Resist Blast

For statically indeterminate systems we calculate an equivalent X_E which allows for loss of stiffness as hinges form (this happens in stages as each hinge forms). See Table 3.5.

Table 3.5: Values of stiffness

Edge Conditions and Loading Diagrams	Stiffness		
	Elastic Stiffness, K_e	Elasto-plastic Stiffness, K_{ep}	Equivalent Stiffness, K_E
⧫⋁⋁⋁⋁⋁⋁⧫	$384EI/5L^4$	-	$384EI/5L^4$
⋁⋁⋁⋁⋁⋁⋁	$8EI/L^4$	-	$8EI/L^4$
⧫⋁⋁⋁⋁⋁⋁	$384EI/L^4$	$384EI/5L^4$	$307EI/L^4$
⧫⋁⋁⋁⋁⋁⋁⧫	$185EI/L^4$	$384EI/5L^4$	$160EI/L^4$

Note: displacement $X_E = w/K_E$

3.5.1.3 *Assumption of Impulsive behaviour*

We have assumed the loading is impulsive. Thus we should check $t_d/T < 0.1$ or so. As mentioned previously, the explosion can be quite close to the structural members (e.g. when basement parking is allowed). This means the load duration (positive phase of the blast) t_d, is usually very small (order of miliseconds). Thus t_d/T is usually small.

An alternative check, which may be more convenient (avoids having to evaluate period T) is to check that t_m/t_d is ≥ 3, where t_m is the time until the maximum deflection is reached ($\approx i/w$) and t_d is the load duration.

This check ensures the load is impulsive and so we can use the impulse equation.

3.6 SUMMARY OF DESIGN PROCEDURE

Assess blast parameters (mostly need impulse).
- Check local effects;
- Assume Impulsive behavior;
- Design as SDOF system: (Check flexure and shear;
- Check assumption of impulsive loading is correct;
- Detail the reinforcement.

Design of Reinforced Concrete Buildings to Resist Blast

3.6.1 Software Available:

SDOF modelling:

SDOF: purchase from Applied Research Associates. (http://www.ara.com/products/software.htm)

SBEDS: distributed free to qualifying parties (must be American citizens).

4

WORKED EXAMPLES

4.1 RC CANTILEVER WALL SUBJECT TO BLAST FROM VBIED.

A cantilever blast wall to be designed to withstand the impulse due to the detonation of 100 kg TNT at ground level at a stand-off of 4 m (bollards centred 2.5 m from wall: assume further 1.5 m to centre of bomb). The wall height is 4 m and it is symmetrically reinforced. Protection category 2 (i.e. lower). Type "B" reinforcement is to be used.

Material strengths (static) are f_{ck} = 35 N/mm^2 and f_{yk} = 500 N/mm^2. The density of the concrete may be taken as 2,500 kg/m^3. Modulus of steel reinforcement E_s = 200 GPa, Mean modulus of concrete E_c = 34 GPa. Take K_{LM} = 0.66

Solution:

Local Effects

The first check we should make is on local effects. This frequently decides the thickness of the wall. Only after

Design of Reinforced Concrete Buildings to Resist Blast

that, is it appropriate to consider the "global" effects (i.e., flexure and shear).

Using the Hader diagram, assess the local effects of a 100 kg TNT charge 4 m from an RC wall. We use a trial and error approach now to guess the appropriate thickness for this wall. The wall will certainly need to be thick. We will guess at least 400 mm.

We must quantify the following parameters: $t/m^{1/3}$, where t is the thickness of the concrete and m is the mass of the charge, and $r/m^{1/3}$, where r is the stand-off.

$t/m^{1/3} = 0.4/(100)^{1/3} = 0.4/4.63 = 0.09$

$r/m^{1/3} = 4.0/(100)^{1/3} = 4.0/4.63 = 0.86$

Now we see where these two points lie on the diagram:

Design of Reinforced Concrete Buildings to Resist Blast

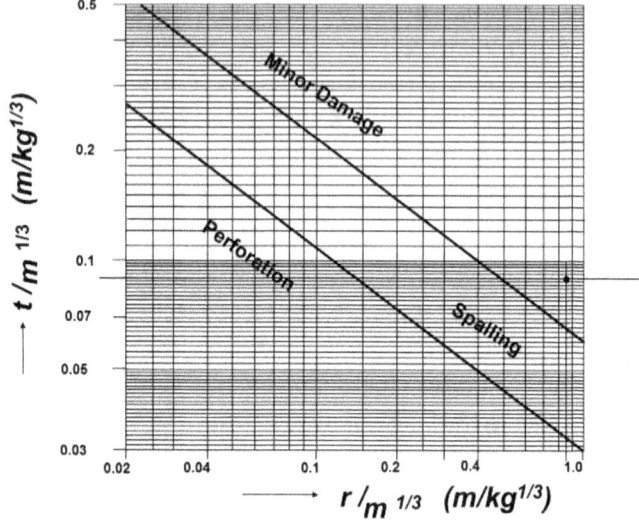

The diagram tells us that there is likely to be only minor damage from this explosion. Thus no spalling is likely and so the wall is appropriate for our use. We can now proceed to the next step. The wall thickness may still increase (but not decrease) if we think necessary, e.g. to improve economy or if the wall fails the direct shear check.

Global Effects: Flexure

- Charge mass, m = 100 kg.
- Stand-off, R = 4 m.
- Scaled distance, $Z = R/m^{1/3} = 4/(100)^{1/3} = 4/4.63 = 0.86$ m.

Refer to "spaghetti chart".

Design of Reinforced Concrete Buildings to Resist Blast

- Reflected impulse,

    $i = 1{,}100 \times (100)^{0.33} = 5{,}100$ kPa.msec

- Duration of positive phase,

    $t_o = 2.1 \times (100)^{0.33} = 9.5$ msec

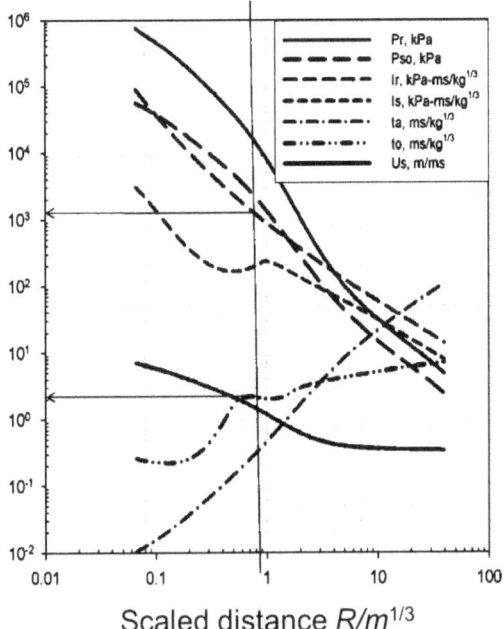

Scaled distance $R/m^{1/3}$

Impulse Equation $i^2 A^2 / 2 K_{LM} M = W(X_m - X_E/2)$

- $i = 5{,}100$ kPa.msec.
- $A = 4$ m².
- $K_{LM} = 0.66$ (Given)

Thus we seek W, X_m, X_E and M.

Design of Reinforced Concrete Buildings to Resist Blast

Consider $b = 1$ m width of wall of height L, lever arm z and a type 2 section (loss of cover in compression zone acceptable). Now we must find the equivalent UDL when the plastic moment capacity (M_p) is reached at the support. Call this UDL W ($= wL$). This is derived below.

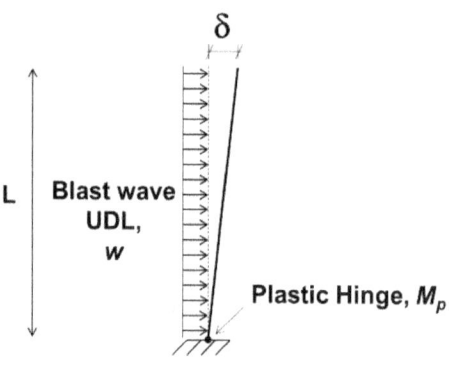

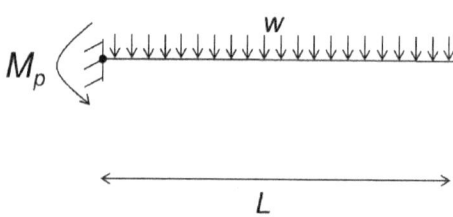

At ultimate, reinforcement yields allowing a hinge to form at the support. The moment capacity of this hinge is M_p. From statics $wL^2/2 = M_p$

Thus $w = 2M_p/L^2$ so $wL = 2M_p/L$

i.e., UDL on cantilever $W = wL = 2M_p/L$

Design of Reinforced Concrete Buildings to Resist Blast

To ensure plastic action we should check that is no brittle failure due to:

- shear (diagonal or direct);
- failure of the concrete as the plastic hinge rotates, i.e. x/d is ≤ 0.25.

For our RC section let's guess z = 300 mm, ($d \approx$ 350 mm, $h \approx$ 400 mm).

$M_p = Tz = A_s f_{yd,dyn} z = r f_{yd,dyn} z^2$ per m width

where $f_{yd,dyn} = 1.2 x f_{yk} = 1.2 \times 500 = 600$ N/mm^2

and $\rho = A_s/bz$ = say 1.6%,

=> $M_p = r f_{yd,dyn} z^2 = 0.016 \times 600 \times 10^6 \times 0.3^2 = 864,000$ Nm

and $W = 2M_p/L = 2(864,000)/L = 432,000$ N = 432 kN

Evaluate X_m (maximum deflection):

Section type 2 (θ = 4 degrees);

From geometry $\tan\theta = \delta/L$

Thus $X_m = L \tan 4° = 4 \times 0.07 = 0.279$ m.

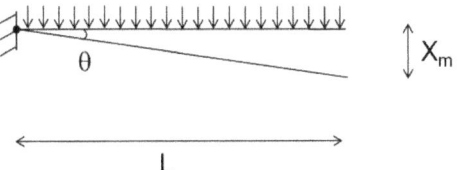

Design of Reinforced Concrete Buildings to Resist Blast

Evaluate X_E (elastic deflection):

Elastic deflection at tip of cantilever subject to UDL, $X_E = wL^4/8EI = WL^3/8EI$. Thus elastic stiffness, $k_e = W/X_E = 8EI/L^3$.

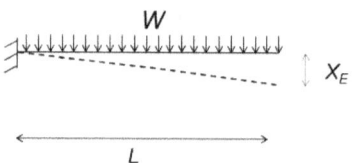

Q: What is the second moment of area I ?

A: We use the transformed section.

For an RC section the second moment of area I is obtained using the following figure, given $\rho \approx 1.6\%$ and $\alpha = E_s/E_c = 200/34 = 5.88$. The following figure gives I.

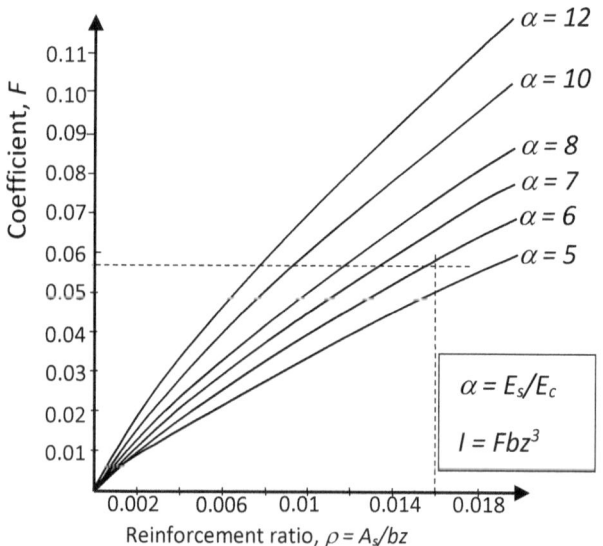

Design of Reinforced Concrete Buildings to Resist Blast

Hence $I = Fbz^3 = 0.057bz^3$

Thus $k_e = 8EI/L^3 = 8 \times 34 \times 10^9 \times 0.057 \times 1 \times 0.3^3/4^3$

= 6,541 kN/m per m width.

Thus $X_E = W/k_e = 432/6,541 = 0.066$ m

Let's review progress: $i^2A^2/2K_{LM}M = W(X_m-X_E/2)$
- $i = 5,100$ kPa.msec = 5,100 N.sec/m².
- $A = 4$ m².
- $W = 432,000$ N.
- $X_m = 0.279$ m.
- $X_E = 0.066$ m.
- $K_{LM} = 0.66$ (given).
- $M = r_c zL = 2,500 \times 4 \times 0.3 = 3,000$ kg.

Substituting

$5,100^2 \times 4^2/(2 \times 0.66 \times 3,000) = 432,000(0.279 - 0.066/2)$

LHS – 105,100

RHS = 432,000(0.28 – 0.06/2) = 106,300;

LHS < RHS => OK!

Use $z = 0.3$ m

$A_s = 0.016 \times 300 \times 1000 = 4,800$ mm²/m

Use B25@100 mm (4,910 mm²) on each face.

Design of Reinforced Concrete Buildings to Resist Blast

Hence overall section thickness, using 25 mm cover and assuming B10 horizontal bars placed outside vertical bars, is h = 25+10+13+300+13+10+25 ≈ 400 mm.

Check assumption of impulsive behavior:

Response time (time until maximum deflection reached), $t_m ≈ i/w$

W = 432 kN per m width.

w = 432/4 = 108 kN/m per m width.

Hence t_m = 5,100/108 = 47.2 msec.

t_m/t_o = 47.2/9.5 = 5 > 3 => impulsive

Check diagonal shear (check at d from support)

UDL w = 108 kN/m per m width. Hence shear at d from support = 108x(4-0.35) = 394.2 kN

Shear stress, $v_{Ed}/b_w z$ = 394,200/(300x1,000)

= 1.31 N/mm^2

Limiting shear stress for cot α = 2.5 (α = 21.8⁰) (using the symbol alpha for the strut angle in EC2 instead of theta used therein) is:

$v_{Rd,max} = 0.4f_{ck}(1-f_{ck}/250))/(\cot \alpha + \tan \alpha) =$

0.4x35(1-35/250)/(2.5 + 0.4) = 4.15 N/mm^2

> 1.31 N/mm^2 => cot α = 2.5

Design of Reinforced Concrete Buildings to Resist Blast

Now $A_{sw}/b_w s = (V_{Ed}/b_w z)/(\cot \alpha \times f_{yd,dyn})$

Thus $A_{sw}/b_w s = (V_{Ed}/b_w z)/(2.5 \times 1.1 f_{yk})$

— where $1.1 f_{yk}$ is the dynamic strength of links.

$A_{sw}/b_w s = 1.31 \times 10^6/(2.5 \times 1.1 \times 500) = 794$ mm²/m²

Check minimum links required by EC2:

Min $A_{sw}/b_w s = (0.08\sqrt{f_{ck}})/f_{yk} = 1 \times 10^6 \times (0.08\sqrt{35})/500 = 947$ mm²/m². Thus the minimum is critical here.

Use links at 150 mm lateral spacing and spacing along span of wall s = 200 mm. The minimum required area of each link is = 947×0.15×0.2 = 28 mm². Therefore use B8 links (50 mm²). Because of low stand-off and support rotation being > 2°, links should have bends s.th angle not less than 135°, i.e.,

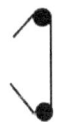

Check Direct Shear:

UDL w = 108 kN/m per m width. Hence shear at the centre of support, $V_{Rd,s}$ = 108×4 = 432 kN

Design of Reinforced Concrete Buildings to Resist Blast

All direct shear should be taken by diagonal reinforcement. We do this as $\theta > 2°$, i.e. take $V_{Rd,max}$ when $\alpha = 45°$ as = 0.

The required area of diagonal bars at $45°$ and spacing b = 150 mm.

$A_{sw,d} = V_{Ed}\, b/(1.1 f_{yk} \sin 45°) =$
$(432 \times 10^3 / (1.1 \times 500 \times \sin 45°)) \times (150/1000) = 166 \text{ mm}^2$

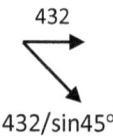

432

432/sin45°

Use B16 bars at 100 mm centres (201 mm² per bar).

Summary:

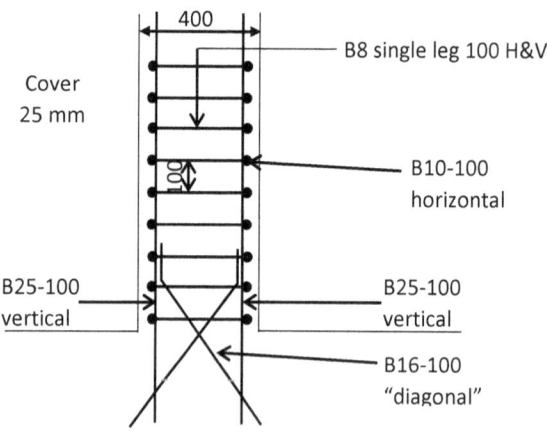

Design of Reinforced Concrete Buildings to Resist Blast

4.2 RC Level 1 Slab Subject to Blast from VBIED in Basement.

A level 1 slab forms a busy lobby area of a large hotel. The hotel has one level of basement car park. A VBIED of 150 kg TNT in the basement is the design attack. The L1 slab is a flat slab of grid 8.4 m x 8.4 m. Basement clear height is 6 m. Columns are square of side 1.2 m.

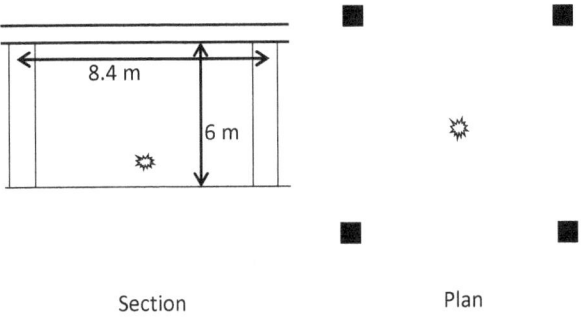

Section   Plan

Using the Hader diagram, find the appropriate slab thickness. Design the reinforcement in a typical bay.

Material strengths (static) are f_{ck} = 35 N/mm^2 and f_{yk} = 500 N/mm^2. The density of the concrete may be taken as 2,500 kg/m^3. Modulus of steel reinforcement E_s = 200 GPa, Mean modulus of concrete E_c = 34 GPa. Take K_{LM} = 0.72 (average of elastic and plastic).

Design of Reinforced Concrete Buildings to Resist Blast

Solution

Local Effects

Using the Hader diagram, assess the local effects of a 150 kg TNT charge 5 m from the RC slab (assume centre of bomb is 1 m off the ground).

We use a trial and error approach now to guess the appropriate thickness for this slab. We will guess 500 mm.

$t/m^{1/3}$, where t is the thickness and m is the mass of the charge; $r/m^{1/3}$, where r is the stand-off.

- $t/m^{1/3} = 0.5/(150)^{1/3} = 0.7/5.3 = 0.13$
- $r/m^{1/3} = 5.0/(150)^{1/3} = 5.0/5.3 = 0.94$

Now we see where these two points lie on the diagram. The diagram tells us that there is likely to be only minor damage from this explosion. Thus no spalling is likely and so the slab is appropriate for L1's use (mass occupancy). We can now proceed to the next step (Global effects).

Flexure

Charge mass, m = 150 kg; Stand-off, R = 5 m; Scaled distance, $Z = R/m^{1/3} = 5/(150)^{1/3} = 5/5.31 = 0.94$ m.

Refer to "spaghetti chart".

Reflected impulse = $1,050 \times (150)^{0.33}$ = 5,579 kPa.msec

Design of Reinforced Concrete Buildings to Resist Blast

Duration of positive phase, $t_o = 2.1 \times (150)^{0.33} = 11.2$ msec

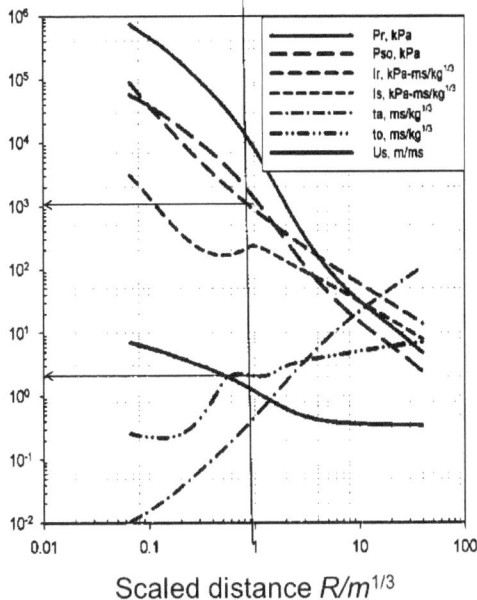

Scaled distance $R/m^{1/3}$

Following VBIED detonation, the wave propagates upwards striking the slab and is reflected back down. Meanwhile the wave also propagates downwards striking the floor below and is then reflected upwards. Thus the waves bounce up and down gaining strength with each reflection. It is reasonable to take the final impulse as 1.1 times the open-air impulse (this can be verified using so-called "hydro-code" software e.g., "Blast X"). If the storey height of the basement is low (say < 3 m or so), an increase of more is likely (up to 1.75 according to Cormie et al. 2009).

Thus we know $i^2A^2/2K_{LM}M = W(X_m - X_E/2)$

- $i = 1.1 \times 5{,}579 = 6{,}137$ kPa.msec;
- $A = 8.4$ m^2 (consider 1 m width);
- $K_{LM} = 0.72$ (Given);
- $M = 2{,}500 \times 0.5 \times 8.4 = 10{,}500$ kg per m width.

Thus we seek W, X_m, and X_E.

Consider $b = 1$ m width of slab of span 8 m and a type 1 section (loss of cover in compression zone is not acceptable). Now we must find the equivalent UDL when the plastic capacity is reached. Call this UDL W (= wL). This is derived below.

Trial section:

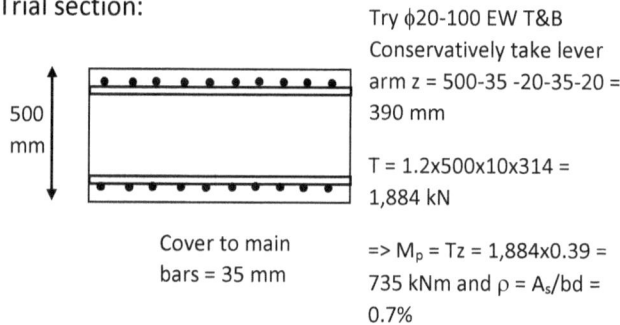

Try φ20-100 EW T&B
Conservatively take lever arm z = 500-35 -20-35-20 = 390 mm

T = 1.2×500×10×314 = 1,884 kN

Cover to main bars = 35 mm

=> M_p = Tz = 1,884×0.39 = 735 kNm and $\rho = A_s/bd$ = 0.7%

Table 3.4 shows that $w = 8(M_{HOG} + M_{SAG})/L^2$

$M_{HOG} = M_{SAG} = M_p$;

From statics $2M_p = wL^2/8$

$\Rightarrow$ $w - 16M_p/L^2 = 16*735/8.4^2 = 166$ kN/m

$\Rightarrow$ $W = wL = 166*8.4 = 1{,}391$ kN

Design of Reinforced Concrete Buildings to Resist Blast

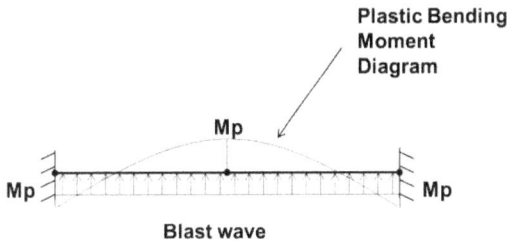

Evaluate X_m (maximum deflection):

Response Limit: $2°$

From geometry $\tan\theta = \delta/L$

Thus $X_m = (L/2)\tan 2° = (8.4/2) \times 0.035 = 0.147$ m.

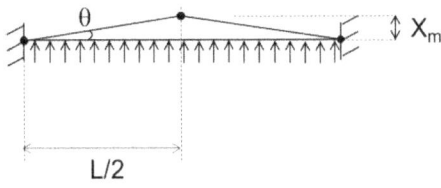

Evaluate X_E (elastic deflection)

Equivalent elastic stiffness K_E of fixed-ended beam under a UDL = $307EI/L^4$ (see Table 3.5).

Now $X_E = w/K_E$

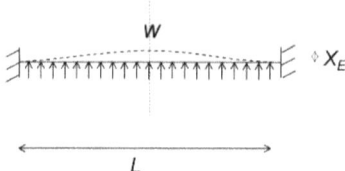

For an RC section the second moment of area I is obtained using the following figure (it is similar to the figure shown previously, except this shows the transformed I for *type 1* sections, i.e. only the cover on the tension side is ignored).

Now $\rho = A_s/bd = 0.7\%$ and $\alpha = E_s/E_c = 200/34 = 5.88$.

Thus $I = 0.027bd^3 = 0.027*1000*445^3 = 2.38*10^9$ mm^4 = 0.00238 m^4.

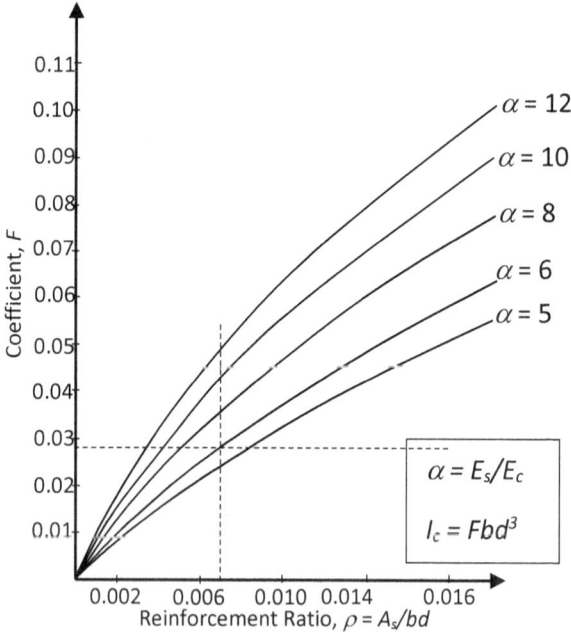

Design of Reinforced Concrete Buildings to Resist Blast

Hence $X_E = w/K_E = w/(307EI/L^4) = wL^4/(307EI)$

$= 166*10^3*8.4^4/(307*34*10^9*0.00238) = 0.033$ m

$i^2A^2/2K_{LM}M = W(X_m - X_E/2)$

LHS $= i^2A^2/2K_{LM}M = 6,137^2*8.4^2 /(2*0.72* 10,500*1000)$
$= 175.6$

RHS $= W(X_m - X_E/2) = 1,391(0.147-0.033/2) = 181.5$

RHS > LHS => OK (conservative to within 5%).

We can iterate again to make this difference smaller. It can be shown that making the slab 485 mm thick will satisfy the impulse equation to within 1%.

Check if assumption of impulsive regime is correct

Response time (time until maximum deflection reached), $t_m \approx i/w$ where $i = 6,137$ kPa.msec and $w = 166$ kN/m per m width. Hence $t_m = i/w = 6,137/166 = 37$ msec.

Now $t_0 = 11.2$ msec, thus $t_m/t_o = 37/11.2 = 3.3 > 3$

=> impulsive

Shear check

As the slab is a flat slab we should check punching (no "beam" shear check is necessary).

Design of Reinforced Concrete Buildings to Resist Blast

We check punching under a load of intensity w =166 kN/m². We estimate the extent of the blast loading at 1 panel.

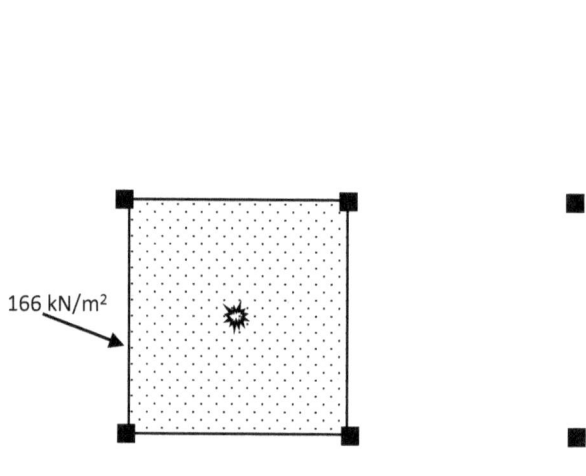

Check punching stress at the column face:

Upward load = 166*8.4²/4 = 2,928 kN

Assume 2 column faces resist the load. Stress v = V/A = 2,928*10³/(1200*2*445) = 2.7 N/mm²

From EC2 $v_{Rd,max}$ for α = 45⁰ and f_{ck} = 35 N/mm² is $v_{Rd,max}$ = *0.4f$_{ck}$(1-f$_{ck}$/250))/(cot α + tan α)*

= 0.4x35(1-35/250)/(1 + 1) = 6.02 N/mm² > 2.7 N/mm²
=> shear at face okay.

Check at 2*d* from column face:

Design of Reinforced Concrete Buildings to Resist Blast

Punching load = 2,928-166×[(2×0.445)²×π+1.2²+1.2*2*0.445*4]/4

= 2,928-166[2.05] = 2,588 kN

Perimeter for 2 sides = 0.5*(2*0.445)*π+1.2*2 = 3.8 m

Stress v_{Ed} = V/A = 2,588*10³/(445*3,800) = 1.53 N/mm².

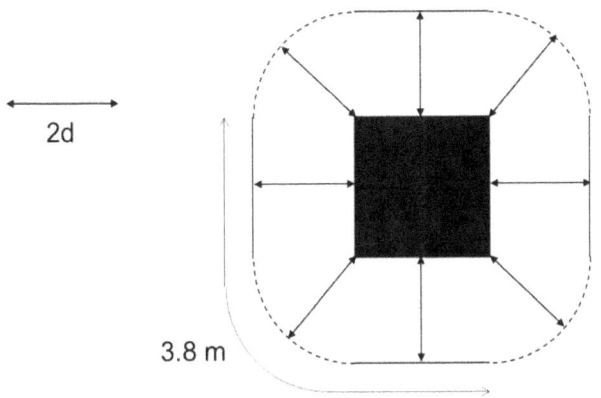

Without links, resistance

$v_{Rd,c}$ = 0.12k(100ρf_{ck})^(1/3) where k = 1 + (200/d)^(0.5)

In our case $\rho = A_s/bd$ = 0.007, f_{ck} = 35 N/mm²

and k = 1 + (200/445)^(0.5) = 1.67

Thus $v_{Rd,c}$ = 0.12*1.67*(0.7*35)^(1/3) = 0.58 N/mm² > 1.53 N/mm². Hence links required.

Design of Reinforced Concrete Buildings to Resist Blast

Calculation of shear reinforcement:

$A_{sw} = (v_{Ed} - 0.75v_{Rd,c})s_r u_1/(1.5f_{ywd,ef})$ per part perimeter.

= (1.53–0.75x0.58)(0.75x445)(3800)

/(1.5x(250+0.25x445)x1.1) = 2,330 mm²

Try 13 mm links => A_{sw} = 2,330/133 = 18 per part perimeter.

First perimeter 0.5d = 0.5x445 = 223 mm from column face.

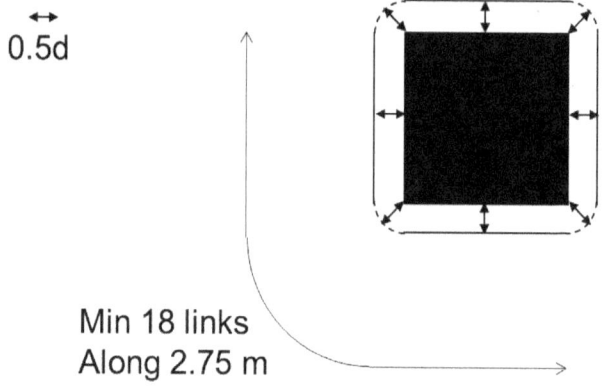

0.5d

Min 18 links
Along 2.75 m

Reinforcing bars for flexure B20@100. Provide links at each 100 mm junction. Thus 12 links per column side => total = 24 on this part perimeter (> 18 => ok).

Next perimeter = 0.75d = 0.75x445 = 334 mm away from this perimeter (1.25d from column). We need 18 links along this part perimeter too. Max spacing allowed along this perimeter = 1.5d = 1.5x445 = 668 mm.

Design of Reinforced Concrete Buildings to Resist Blast

Next perimeter is 0.75*d* away (i.e. 2d from column). Length if this part perimeter = 3.8 m. We need min 18 links over this length, i.e. 3.8/18 = 0.21 m crs.

Use 200 mm for convenience.

Next perimeter 0.75*d* away (i.e. 2.75*d* from column). Length = 4.3 m => spacing = 4.3/18 = 0.24 m

Say 200 mm for convenience.

Restraint to compression bars: EC2 says:

Longitudinal compression reinforcement should be within 150 mm of a restrained bar and held by transverse reinforcement with spacing not greater than 15 times diameter of longitudinal bar.

This means our link spacing cannot exceed about 300 mm. These links are required throughout the span.

Summary:

Perimeter Distance from column (m)	Part-Perimeter length (m)	Spacing of links along perimeter (m)	Spacing used (m)
0.5*d* = 0.22	2.75	0.15	0.10
1.25*d* = 0.56	3.27	0.18	0.10
2.0*d* = 0.89	3.80	0.21	0.20
2.75*d* = 1.22	4.30	0.24	0.20
3.5*d* = 1.56	4.85	0.27	0.20
4.25*d* = 1.89	5.37	0.30	0.30

Design of Reinforced Concrete Buildings to Resist Blast

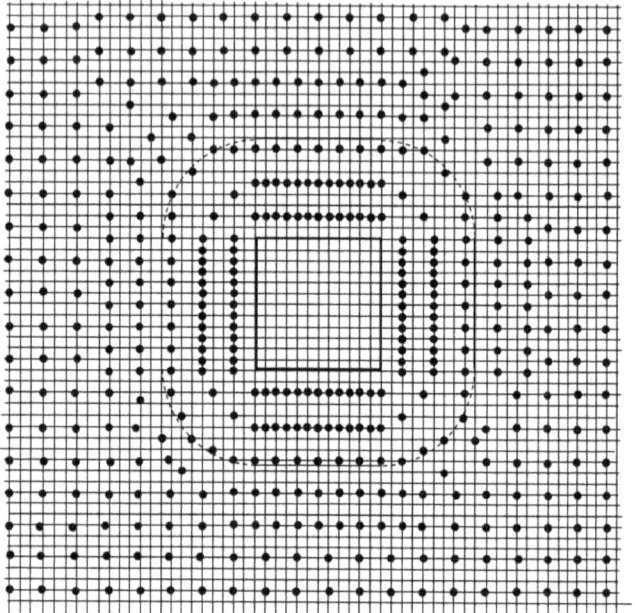

Note: Link Detailing:

Link detailing: Outer layers of flexural bars should be restrained. Thus links must enclose outer bars.

Alternative solution: use thinner slab (about the middle of "spalling" part of Hader diagram) with top steel anti-spalling plate; cast-in bolts at 500 mm crs put in place using a template. Top plate has holes to line-up with bolts. Plate later made continuous by welding. Clearly this is an expensive and slow solution.

4.3 RC COLUMN SUBJECT TO BLAST FROM VBIED.

A column in the basement car park of a multi-storey building is to withstand the impulse due to the detonation of 100 kg TNT 1 m off the floor at a stand-off of 1.5 m. The basement clear height 6 m.

Check the section shown below. The axial load corresponding to DL+0.5LL is 5,000 kN. Section type 1 (i.e. 2 degree rotation acceptable). Material strengths (static) are f_{ck} = 50 N/mm² and f_{yk} = 500 N/mm².

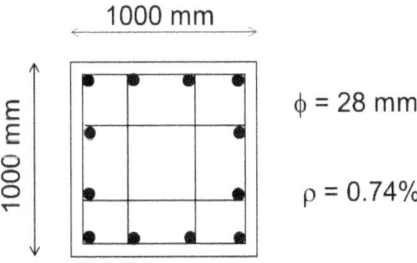

Using dynamic material strengths, M_p is calculated to be 4,071 kNm (including effect of axial load) and x/d = 0.25. Effective d = 730 mm.

Design of Reinforced Concrete Buildings to Resist Blast

Modulus of steel reinforcement E_s = 200 GPa, Mean modulus of concrete E_c = 34 GPa. Reinforcement is type "B". Take K_{LM} = 0.72.

Scenario:

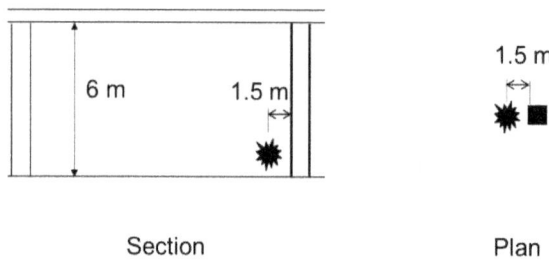

Section   Plan

Note: reasonable to guess VBIED can get c.o.g. of bomb no closer than 1.5 m

Solution

Local Effects:

We must quantify the following parameters:

- $t/m^{1/3}$, where t is the thickness of the concrete and m is the mass of the charge.
- $r/m^{1/3}$, where r is the stand-off.

$t/m^{1/3}$ = $1.0/(100)^{1/3}$ = $1.0/4.63$ = 0.22 and

$r/m^{1/3}$ = $1.5/(100)^{1/3}$ = $1.5/4.63$ = 0.32

Now we see where these two points lie on the diagram

Design of Reinforced Concrete Buildings to Resist Blast

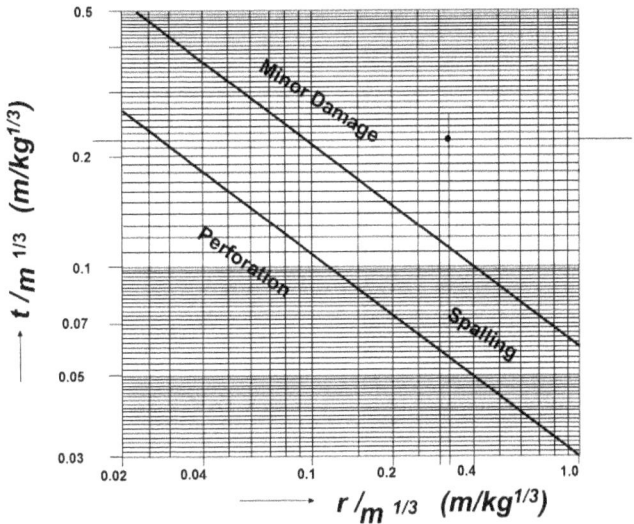

The diagram tells us that there is likely to be only minor local damage from this explosion. We can now proceed to the next step.

Blast Parameters

Charge mass, $m = 100$ kg; Stand-off, $R = 1.5$ m; Scaled distance, $Z = R/m^{1/3} = 1.5/(100)^{1/3} = 0.32$ m. Refer to "spaghetti chart".

Thus Reflected impulse = $5{,}000 \times (100)^{0.33}$

= 23,208 kPa.msec

Duration of positive phase, $t_o = 0.5 \times (100)^{0.33} = 2.3$ msec.

Design of Reinforced Concrete Buildings to Resist Blast

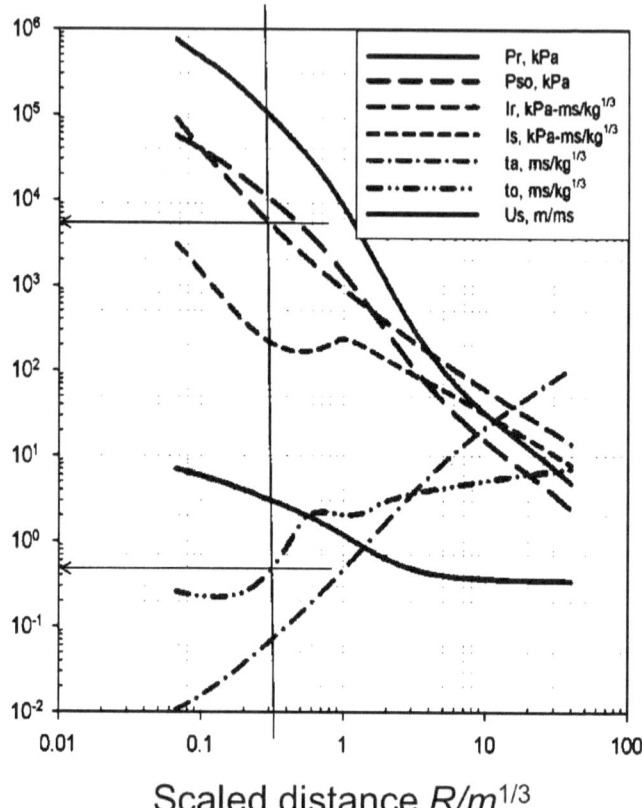

Scaled distance $R/m^{1/3}$

Impulse Equation:

We require energy in = energy absorbed,

i.e., $i^2A^2/2K_{LM}M = W(X_m - X_E/2)$

where i = 23,208 kPa.msec, A = 6*1 = 6 m², K_{LM} = 0.72, M = 2,500*1*1*6 = 15,000 kg

Thus LHS = $i^2A^2/2K_{LM}M$

= (23,208*6)²/(2*0.72*15,000) = 897,685 Nm.

Thus we seek W, X_m, and X_E.

When w is applied, reinforcement yields allowing three hinges to form.

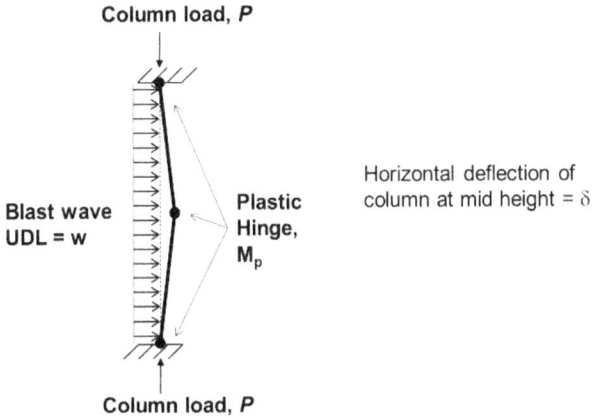

From Table 3.4, $w = 8(M_{HOG}+M_{SAG})/L^2 = 16M_p/L^2$

$\Rightarrow w = 16*4071/6^2 = 1809$ kN/m

$\Rightarrow W = 1809*6 = 10,856$ kN

Maximum Deflection, X_m:

We'll take the response limit as 2 degrees at upper and lower supports for now.

Thus $X_m = (6/2) \tan 2° = 0.104$ m.

Design of Reinforced Concrete Buildings to Resist Blast

Deflection corresponding to elastic limit, X_E:

Equivalent elastic stiffness (from Table 3.15) $K_E = 307EI/L^4$.
And $X_E = w/K_E$

Again we use the transformed section for I. For an RC section the second moment of area I is obtained using the same figure as in the previous example, given $\rho = 0.33\%$ on each face (taking 4 bars at face only) and $\alpha = E_s/E_c = 200/34 = 5.88$.

Hence $I = Fbd^3 = 0.014bd^3 = 0.014*1*0.73^3 = 5.45*10^{-3}$ m^4

Thus $X_E = w/K_E = wL^4/307EI$

$= 1809*10^3*6^4/(307*34*10^9*5.45*10^{-3}) = 0.041$ m

Impulse Equation: $RHS = W(X_m - X_E/2)$

Where $W = 8,520$ kN; $X_m = 0.104$ m; $X_E = 0.041$ m.

(Note: Ductility ratio $\mu = X_m/X_E = 2.5 < 3$ as required by DoD 2008 => ok).

$W(X_m - X_E/2) = 10,856*10^3*(0.104 - 0.041/2) = 906,000$ Nm

Compare with $LHS = 897,000$ Nm

RHS bigger and within 1% => ok

Check P-delta effect:

Axial load = 5,000 kN; $X_m = 0.104$ m.

Hence $P\delta = 5,000*0.104 = 520$ kNm

Design of Reinforced Concrete Buildings to Resist Blast

Now M_p = 4,071 kNm so $P\delta/M_p$ = 12.8% which is lower than the 15% allowed => ok

Check if assumption of impulsive regime is correct

Response time (time until maximum deflection reached), $t_m \approx i/w$ and t_o = 2.3 msec.

W = 10,856 kN; w = 10,856x10³/6 = 1,809 kN/m.

Hence t_m = 23,206/1,809 = 12.8 msec.

Thus t_m/t_o = 12.8/2.3 = 5.6 > 3 => impulsive.

Check diagonal shear (check at *d* from support)

UDL w = 1,809 kN/m.

Hence shear at *d* from support = 1,809x(3-0.73)

= 4,106 kN

Shear stress, $V_{Ed}/b_w z$ = 4,106,000/(1000*0.9*730)

= 6.2 N/mm²

Limiting shear stress for cot α = 2.14 (α = 25⁰) is,

$v_{Rd,max}$ = *0.4f$_{ck}$(1-f$_{ck}$/250))/(cot α + tan α)* =

0.4x50(1-50/250)/(2.14 + 0.47) = 6.13 N/mm²

Allow for axial load enhancement => capacity = 1.1*6.13 = 6.74 N/mm² > 6.2 N/mm²

=> cot α = 2.14

Now $A_{sw}/b_w s = (V_{Ed}/b_w z)/(\cot \alpha \times f_{yd,dyn})$

Thus $A_{sw}/b_w s = (V_{Ed}/b_w z)/(2.14 \times 1.1 f_{yk})$

where $1.1 f_{yk}$ is the dynamic strength of links.

$A_{sw}/s = 6.2 \times 1000/(2.14 \times 1.1 \times 500) = 5.26$

Try B13 links.

2 pairs => $A_{sw} = 133 \times 4 = 532$ mm²

=> spacing, s = 532/5.26 = 101 say 100 mm.

Therefore use pairs of B13 links @ 100 mm.

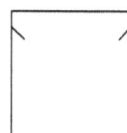

Check direct shear

UDL w = 1,809 kN/m.

Hence shear at the centre of support, $V_{Rd,s} = 1,809 \times 3 = 5,427$ kN

$v_{Ed,z} = 5427,000/(1000 \times 0.9 \times 730) = 8.2$ N/mm²

$v_{Rd,max} = 0.4 f_{ck}(1 - f_{ck}/250)/(\cot \alpha + \tan \alpha) =$

$0.4 \times 50(1 - 50/250)/(1 + 1) = 8.0$ N/mm²

Thus $v_{Rd,max}$ when $\alpha = 45°$ is 8.0 N/mm².

Allowing axial load enhancement => allowable stress = $1.1 \times 8.0 = 8.8$ N/mm².

Design of Reinforced Concrete Buildings to Resist Blast

> required stress of 8.2 N/mm^2 => ok.

As we are close to the limit, add 5 mm thick steel jacket (full-height) to shield concrete. Use an architectural covering to clad the column of 150 mm if possible.

It is good practice to ensure the column is "short" over two storeys in the event lateral resistance at slab level is lost and to provide the column with seismic links (see Ch 9 of ACI-318) over the full height.

If a column fails the check for blast resistance, then there are several options:

- Column should be protected by cladding to increase stand-off.
- Security measures may be adopted to prevent public vehicles in the basement car park.

Otherwise the column should be assumed to be lost and progressive collapse resistance ensured. This is the subject of the next chapter.

5

PROGRESSIVE COLLAPSE

A structure has good resistance to progressive collapse if it can prevent a *local* collapse becoming a *global* collapse. This resistance is also known as "structural integrity" or "robustness" or "redundancy". A statically determinate structure has poor progressive collapse resistance. By definition statically determinate implies the structure is only one step away from being unstable.

Consider a simply supported beam:

Suppose support at B is lost for some reason. Then, no matter how strong the beam, or support A is, the structure is **unstable** and so will collapse. There is a simple rule to identify statically determinate structures: there are just enough members, constraints between members, and supports, to preserve its shape and keep it stable and at rest under load. Most *insitu* concrete structures are statically indeterminate. However it is not enough to ensure that the structure is statically indeterminate. Adequate **strength** must be provided to ensure the structure is stable in its new configuration. If

Design of Reinforced Concrete Buildings to Resist Blast

any one member, constraint, or support is removed, the structure collapses, i.e., undergoes a *major* change of shape.

Consider a propped cantilever beam:

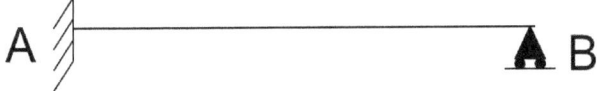

Suppose again support at B is lost. It can remain **stable** in this new configuration *only* if the beam can resist the cantilever moments and shears, and the support at A can resist the cantilever moment and shear, i.e., **if it is strong enough**. Only sufficient strength is required so that warning of collapse is given. The structure should be stable after an accident for long enough time to evacuate. Accident safety factors apply, e.g., EC2: $DL+0.5LL$ with γ_m (for concrete and steel = 1.2 and 1.0 respectively). Serviceability criteria (e.g., deflections) do not apply.

Note: When the structure is statically determinate there is no alternative load path to the supports. With each redundancy another possible load path is available. Thus the more redundancies then, at least potentially, the safer the structure.

After an accident how can an RC beam carry load? See Fig 5.1.

Design of Reinforced Concrete Buildings to Resist Blast

1. Beam action: remaining RC beam tries to support load spanning to its new supports.

2. Catenary action: concrete assumed failed; only resistance of continuous reinforcement remains.

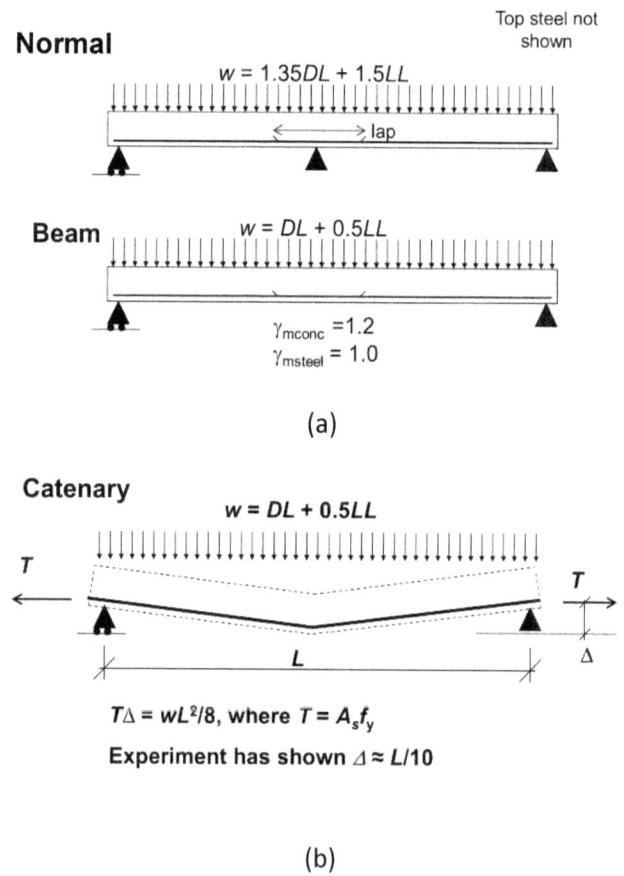

Fig. 5.1: Beam action (a); Catenary action (b)

Catenary action resists the load. This depends on the **detailing** of the bottom reinforcement of the beam. Members may be required to act in "unexpected" ways to resist load. The entire building may act as a **vierendeel truss** (truss without diagonals) to carry load after a column is removed. (Structure tries to arch over missing column.) Often other means of carrying the load to the foundations can be used. See Fig. 5.2 where two cases of column loss are considered.

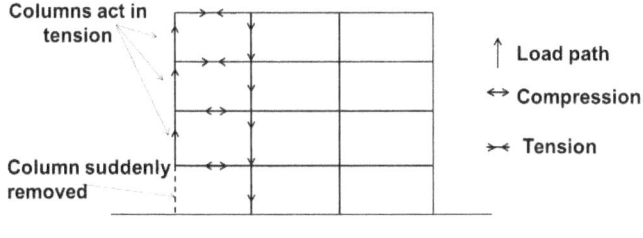

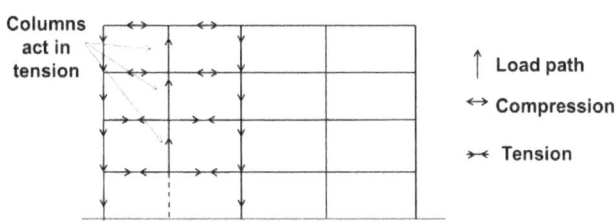

Fig. 5.2: Vierendeel action

We will now look at the way that EC2 ensures that good resistance to progressive collapse is provided in the case of conventional accidental damage (e.g. gas explosion) rather than terrorist attack, which of course this code does not cover.

5.1 Review of EC2

EC2 gives 3 *alternative* methods for design against progressive collapse:

1. "Ties": a "deemed to satisfy" method. This method is one usually adopted in normal design. OR

2. "Alternative Load Path": Explicitly consider removal of one column on any storey and ensure collapsed area < lesser of 15% or 70 m² floor. OR

3. "Key element": Design all members as key elements. A key element is one whose collapse would lead to unacceptable consequences, i.e., above limits exceeded.

1.   Ties

The types of horizontal tie are shown in Fig. 5.3. The code recommends the area of reinforcement for each tie. Tie strength required (T in Fig. 5.4 below):

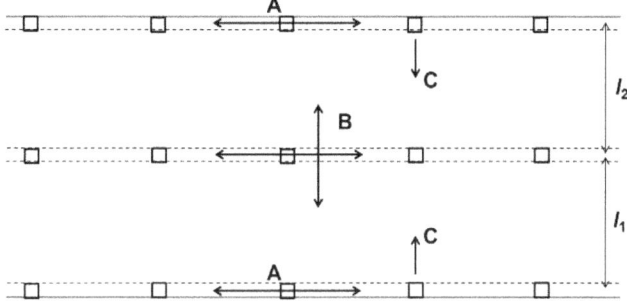

A - peripheral tie; B - internal tie; C - horizontal column or wall tie

Fig. 5.3 Types of horizontal tie

Lesser of $(4n + 20)$ kN/m, where n is the number of stories; and 60 kN/m.

Note: here the implication is that the higher the building, the worse the problem.

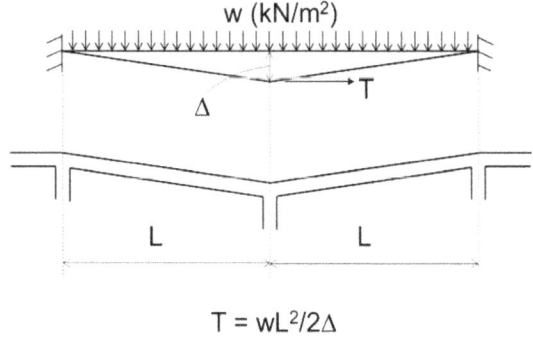

$T = wL^2/2\Delta$

Fig. 5.4: Catenary model of EC2

Substituting typical values of $L = 5$ m, $w = 5$ kN/m² and $\Delta = L/5$ (verified by experiment), gives $T = 62.5$ kN/m.

Note 1: suppose provide B10@500 fully lapped reinforcement => $T = 78.5 \times 2 \times 500 = 78.5$ kN. (Recall: this is *not* additional reinforcement, as the beam is not using flexure to resist the load).

Note 2: $w = 5$ kN/m² corresponds to the s/wt of a 200 mm slab.

Vertical ties are also specified by the code although the minimum column reinforcement is usually sufficient to meet this requirement. The column thus has some tensile resistance.

2. **Alternative Load Paths**

Each column is removed, one at a time, and an alternative path for the load to be transmitted to the foundations found.

3. **Key Elements**

Design to resist 34 kPa (no safety factors) applied in turn horizontally and vertically (assuming accident load acting simultaneously);

5.2 Design against terrorist attack

As previously mentioned, EC2 does not include design against terrorist attack. The method usually used is known as the "Enhanced Local Resistance" method. Thus usually columns and floors exposed to attack are **hardened** to protect them. Reliance is only put on the progressive collapse resistance if such resistance can be proven by calculation.

Modifications for blast design:

- blast causes substantially greater effect than that covered by the "key element" load 34 kPa;

- the dynamic effect of member removal is to be considered unless sufficient hardening provided. Note the impact factor may be as low as 1.4 according to Tian (2011).

5.3 Summary:

- Provide **at least** ties required by code (Providing ties crucial as prevents debris build up from non-exposed parts of structure).

- Hardening key-elements to ensure a blast will not remove them.

Design of Reinforced Concrete Buildings to Resist Blast

- Prevent the blast wave from entering the structure as much as possible. Cladding should be blast resistant.

- Consider gravity load during blast event as DL + 0.5LL.

- Otherwise, Progressive Collapse resistance must be proven by calculation.

- Then imagine configuration of building with member removed by blast.

- Ensure alternative load paths are available.

- Non-structural elements unaffected by blast, e.g., masonry walls, can help. Changing the structural system from one-way to two-way should also be considered.

- Some software is helpful (e.g., Extreme Loading of Structures (www.extremeloading.com)).

6

Façade Design

This section deals with the design of glass and their surrounding walls. The façade should be able to resist the blast wave, otherwise there is a risk of high casualties and much more serious damage to the structure.

Note: The design charge recommended by Singapore's MHA is often found to be small enough (as long as the stand-off from the perimeter is reasonably high, say 10 m or so) that the risk of *structural* damage as a result of façade failure is low.

6.1 Sequence of blast effects-VBIED outside building:

1. Blast wave breaks windows. Perimeter columns/wall blown in. (Fig. 6.1 (a)).

2. Blast wave forces floors upwards. (Fig. 6.1 (b)).

3. Blast wave surrounds structure. Downward pressure exerted on roof.

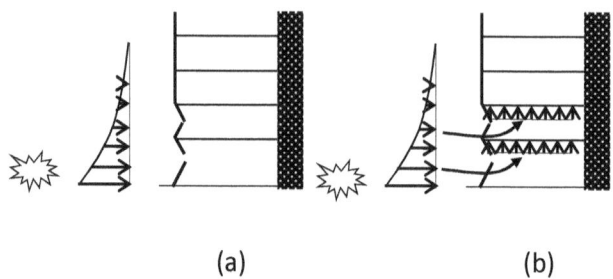

Fig. 6.1: Blast wave enters if façade fails

If the façade is improperly designed, then as well as the direct effect on perimeter columns, the blast wave can penetrate the building, lifting slabs and potentially removing lateral restraint to columns and causing their failure. Even if the structure is not damaged by the wave entering the building, it can result in expensive non-structural damage.

6.2 MASONRY WALLS

The normal wall used for the façade of conventional buildings is un-reinforced masonry. However this is not suitable for blast resistant design as the un-reinforced masonry is brittle. Instead, RC walls spanning vertically (so they do not act to increase the load on columns) are preferred. On the other hand, RC walls often have a large impact on the programme as they are slow to construct. Reinforced masonry is acceptable for smaller impulses.

US DoD (2007) recommends that: "Un-reinforced masonry walls are prohibited for the exterior walls of new buildings".

6.3 Glazing

There are several available glass types:

- Annealed Glass;
- Toughened Glass;
- Security Glass (with wire mesh);
- Laminated Glass.

Normally, non-hardened windows consist of annealed glass of about 6 mm thickness. For blast resisting applications, it will be seen that only laminated glass is suitable.

6.3.1 Annealed Glass

Such glass is typically weak and breaks (at around 14 kPa) into sharp shards which are likely to cause serious injury. This weakness is the main reason strengthening a façade to resist blast can be so costly.

Glass is an amorphous (non-crystalline) material: atoms are in disordered state. It is made by heating silica sand (SiO_2). Sometimes other ingredients added to achieve better properties, e.g. boron oxide to get borosilicate glass. Window glass is "floated" by pouring it into a tin bath while molten, giving "float glass". The cooling rate is often controlled by placing the tin into an oven ("annealing").

Unlike most engineering materials, the properties of the surface of the glass are important, rather than those of the inside. Even very small scratches on the surface can weaken the glass substantially. Thus only small surface tensions are allowed. This low tensile strength and brittleness makes conventional glass unsuitable for structural use. In addition, shards of conventional annealed glass are sharp and cause serious injury.

6.3.2 Toughened Glass

The solution for static loads is a heat treatment process which results in "toughened glass". This usually at least doubles the strength. The section of glass is prestressed by forcing the outer layers into compression. This residual stress may be up to 70 N/mm². Thus when the subsequent load acts it must overcome the compression first before the surface is forced into tension.

The stresses that result are as follows: (see Fig. 6.2)

Design of Reinforced Concrete Buildings to Resist Blast

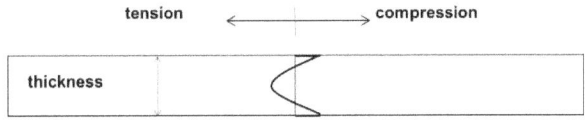

Fig. 6.2: Toughened Glass

This is done by heating then suddenly cooling the surface (using water jets). The surface contracts more than the still-soft inside.

To see this, imagine there are just three layers of glass: two outside and one interior (Fig. 6.3 (a)). The outer layers cool quickly and try to shrink. There is little resistance to their shrinkage as the interior is still soft. Later on, the inside layer cools. It tries to move relative to the outer (now hardened) layers (Fig. 6.3 (b)). They resist this, so they go into compression and impose an equal and opposite tension on the inner layers.

(a)

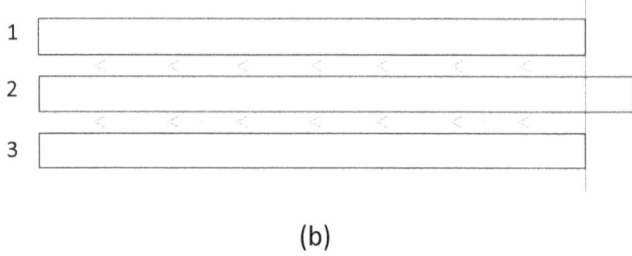

(b)

Fig. 6.3: Manufacturing Principle.

When broken toughened glass usually shatters into small blunt square pieces. In a blast these would move at high velocity resulting in injuries. This makes it **unsuitable** for blast resisting applications. It may also be unsuitable in static load applications because of the problem of spontaneous breakage. There are a number of causes of such spontaneous breakage:

- Inclusions (such as nickel-sulphide) that cause rupturing when heated;
- Differential thermal expansion from unusual shading patterns;
- Edge defects from manufacturing or handling;
- Excessive loading;
- Missile impact (e.g. wind-borne gravel from roof ballast);
- Contact with metal framing due to poor installation practice.

Design of Reinforced Concrete Buildings to Resist Blast

In safety-critical applications, e.g. skylights or safety barriers, it is recommended that tempered (i.e. toughened) glass should be replaced by **laminated** glass due to concerns of spontaneous breakage.

6.3.3 Laminated Glass

Laminated glass is formed by bonding two or more layers of glass, together with resin (often PVB (polyvinyl butyral resin)), using heat and pressure.

PVB when loaded at high rates is very ductile and has significant tensile strength. It bonds well to glass, so on breaking holds the fragments. After the glass breaks the PVB acts as a catenary to withstand the applied load. See Fig. 6.4.

Thus laminated glass must be well anchored into the supports to allow it to behave as a catenary. A "bite" of 20 mm with silicone, to act as a glue and reduce the stress concentrations, is often specified for blast resistance.

In blast applications it is common to assume the laminated glass can deflect up to 200 mm. (Thus in double glazing, at least the inner lite should be laminated.)

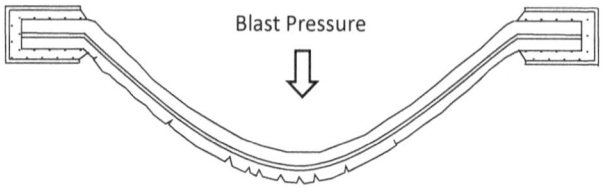

Fig. 6.4: PVB allows the pane to act as a catenary

The vast majority of the resistance comes from the PVB layer. Fig. 6.5 shows a typical curve of energy absorption. It can be seen that most of the energy is absorbed by the catenary action of the glass.

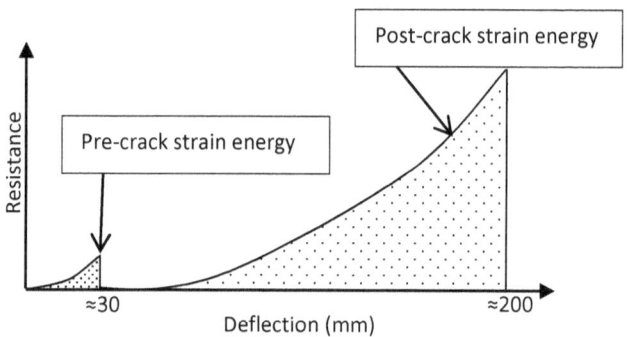

Fig. 6.5: Typical graph of resistance

Toughened glass is not preferred for blast applications as the glass breaks into small blunt cubes which travel at high speed and contribute to injuries. Security Glass, which incorporates a mesh, is not preferred for similar reasons. A popular retrofit method for increasing the blast resistance of existing glass is to apply security film.

6.4 OVERALL CLADDING BEHAVIOUR

To improve blast behaviour, the cladding system should span <u>vertically</u> from floor to floor. This is to avoid placing additional load on the column (otherwise the cladding acts as a "collector" of the lateral load, which is then placed on the column). See Fig. 6.6.

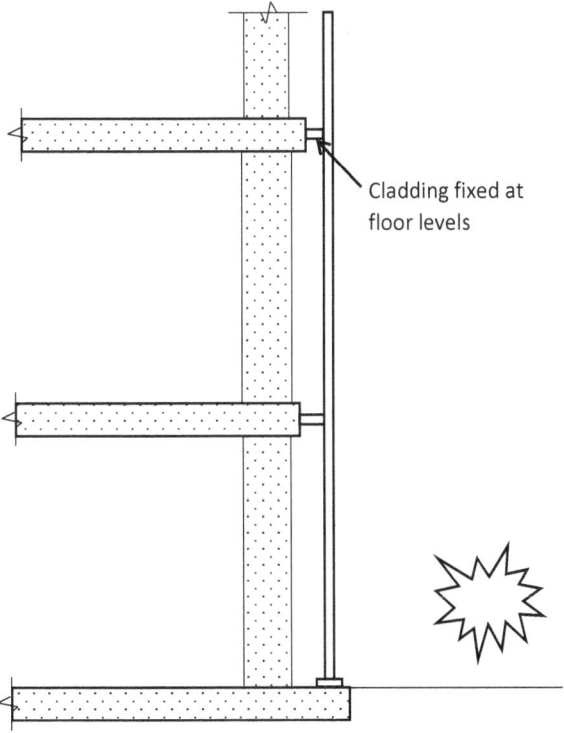

Fig. 6.6: Cladding should span vertically

Design of Reinforced Concrete Buildings to Resist Blast

The design of glazing (and their frames) should be carried out using recognized software (e.g., WinGard by ARA inc.). There are several design specifications internationally recognized as measures of glazing performance. For example, the GSA (2003) criteria rate the performance from 1 to 5.

For reflected impulses of up to about 1,400 kPa.msec, the glazing can be rigidly attached to the frame. For reflected impulses of up to 4,200 kPa.msec the window glass should be fixed to the frame using **Energy Absorbing Units** (EAUs). The latter are steel spring-like devices that normally reside within the frame and are extended once the window is subject to a blast (e.g., those produced by Arpal Aluminium Ltd). They are always used in conjunction with laminated glass. In addition, the forces delivered to the structure are substantially reduced.

For even higher blast impulses vertical stainless steel cables are placed behind laminated glass. This is the **cable catcher** system. Again the cables are always used in conjunction with laminated glass. When the blast wave strikes the glass is pushed back against the cables, but remains in one piece. (See Fig. 6.7).

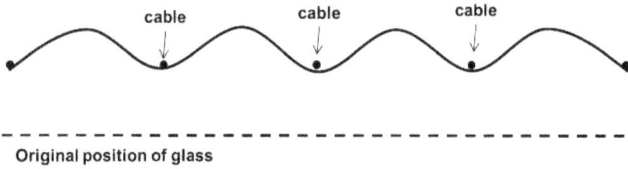

Fig. 6.7: cable catcher system

Note: Glazing has been verified by full-scale blast testing, for a maximum impulse of about 5,250 kPa.msec. Thus for values of impulse above this limit, there is no *verified* solution. This corresponds to the reflected impulse from a 250 kg detonation at about 6.3 m.

6.4.1 Design of frame

Frequently aluminium (Grade 6061), but for higher blast loads there should be a steel insert inside the frame members as aluminium is prone to local buckling and high deflections as E is low (typically 70 kN/mm^2). The limit for deflection is span/60.

7

DERIVATIONS

7.1 SDOF MODEL: DERIVATION OF K_L, K_M, K_{LM}, K_S

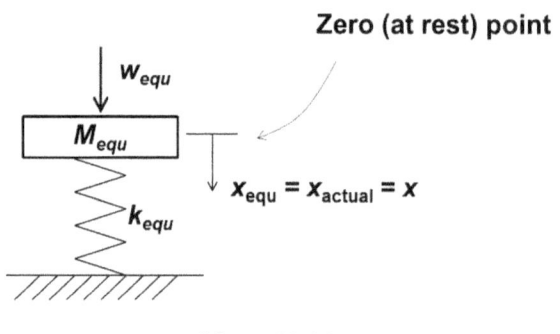

$$M_{equ} = K_M M_{actual}$$

$$w_{equ} = K_L w_{actual}$$

Note: the displacement is the same in the actual and SDOF systems.

Real systems' equation of motion: $M\ddot{x} + kx = w$

=> SDOF equation of motion: $K_M M\ddot{x} + K_L kx = K_L w$

i.e. $K_{LM} M\ddot{x} + kx = w$ where $K_{LM} = K_M/K_L$

Consider a simply supported beam behaving elastically.

Design of Reinforced Concrete Buildings to Resist Blast

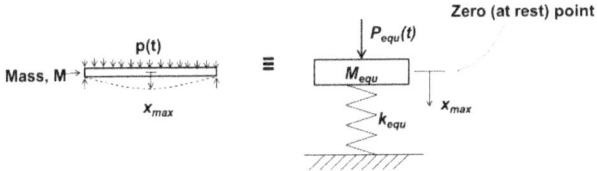

The equivalent system, shown on the right hand side, has the same maximum displacement, x_{max} and maximum initial velocity $\dot{x}_o$ as the real structure.

However other quantities are different:

i.e., $P_{equ} = K_L P$; $M_{equ} = K_M M$; $k_{equ} = K_S k$, where $K_s = K_L$.

The load is applied dynamically (varies with time). The deflected shape of a s/s beam is as follows:

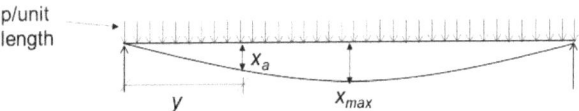

- $x_a = py(L^3 - 2Ly^2 + y^3)/24EI$ ....(1)
- $x_{max} = 5pL^4/384EI$ ....(2)

Thus $p/EI = 384 x_{max}/5L^4$ .... From (2)

And $x_a = 16(L^3 y - 2Ly^3 + y^4) x_{max}/5L^4$... Sub for p/EI into (1)

Total work done by load = $\sum$ work done by components of beam (over length dy).

$\Rightarrow$ Work $= \int 0.5\, p\, x_a\, dy$ .... from $y = 0$ to $y = L$

$= \int 0.5\cdot 16p(L^3 y - 2Ly^3 + y^4) x_{max}\, dy/5L^4$

$= 16p(L^3 y^2/2 - 2Ly^4/4 + y^5/5) x_{max}/10L^4$

Design of Reinforced Concrete Buildings to Resist Blast

$= 16p(L^5/2 - 2L^5/4 + L^5/5)x_{max}/10L^4$

$= 16pLx_{max}/50 = 16Px_{max}/50$

Work done by the equivalent system $= 0.5P_{equ}x_{equmax}$

But $x_{equmax} = x_{max}$. Thus $16Px_{max}/50 = 0.5P_{equ}x_{max}$

$\Rightarrow P_{equ} = 16P/25$

i.e., $K_L = P_{equ}/P = 16/25 = 0.64$

The load and stiffness factors must be equal, i.e. $K_S = 0.64$. This is because when a static load is applied $\ddot{x} = 0$ and so the term $m\ddot{x}$ in the equation of motion is also 0. Thus the equation of motion becomes $kx = F$. And thus the modification to k is the same as to F.

Now, from previous work:

$x_a = 16(L^3y - 2Ly^3 + y^4)x_{max}/5L^4$

Differentiate w.r.t. time:

$\Rightarrow \dot{x}_a = 16(L^3y - 2Ly^3 + y^4)\dot{x}_{max}/5L^4$

$\Rightarrow$ Kinetic Energy of system, $KE = \int 0.5(mdy)\dot{x}_a^2$
where m is the mass per unit length

$\Rightarrow KE = \int 0.5(mdy)[16(L^3y - 2Ly^3 + y^4)\dot{x}_{max}/5L^4]^2$

$= 0.25M\dot{x}_{max}^2$

Now Kinetic Energy of equivalent system

Design of Reinforced Concrete Buildings to Resist Blast

$= 0.5 M_{equ} \dot{x}_{equ}^2$ And $\dot{x}_{equ} = \dot{x}_{max}$

$\Rightarrow M_{equ} = 0.5M$, i.e., $K_M = M_{equ}/M = 0.5$

Thus $K_{LM} = K_M/K_L = 25/32 =$ **0.78**

7.2 FORMULAE FOR RESISTANCE

As an example, let's consider a fixed-ended beam under a UDL. Firstly, assume a collapse "mechanism".

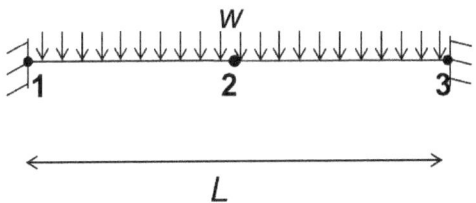

Next, we observe that the maximum static bending moment under a load w is $wL^2/8$.

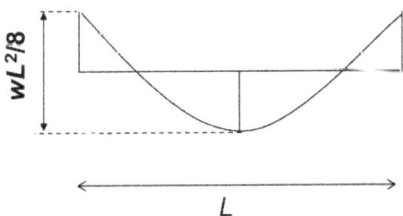

Here we need 3 hinges to collapse this structure. Hence 1, 2 and 3. The bending moment at the two support hinges is M_{HOG} while that at mid-span is M_{SAG}.

Next, we observe that the maximum static bending moment under a load w is $wL^2/8$.

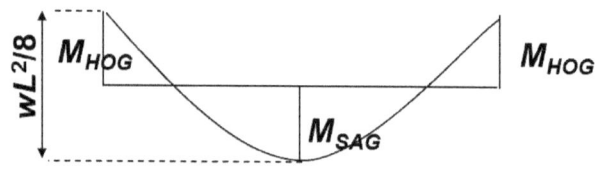

$M_{HOG} + M_{SAG} = wL^2/8$,

Thus resistance, $w = 8(M_{HOG}+M_{SAG})/L^2$

7.3 Formulae for Equivalent Elastic Stiffness

As an example, let's consider a fixed-ended beam under a UDL.

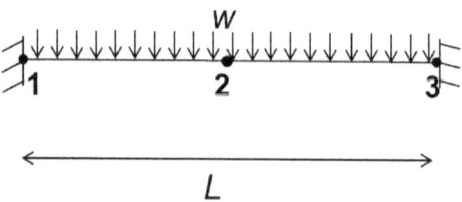

From the previous derivation we know that the plastic capacity is $w_p = 16M_p/L^2$ (assuming $M_{sag} = M_{hog}$).

Design of Reinforced Concrete Buildings to Resist Blast

We calculate an equivalent value of X_E which allows for loss of stiffness as hinges form (this happens in stages as each hinge forms).

We know the moment at the support before the formation of any hinges is $w_{elastic}L^2/12$. The stiffness (load/mid-span deflection) is $K_e = 384EI/L^4$.

As the load increases two hinges at the supports form simultaneously (from symmetry). For subsequent loading (w_{ep}) the beam behaves as simply supported. The mid-span deflection of a simply supported beam is $5/384\, w_{ep}L^4/EI$, thus its stiffness is $K_{ep} = w_{ep}/\delta = 384EI/5L^4$.

When all three hinges eventually form the beam loses all stiffness.

Equating the support moments in the fully elastic case to those when the mechanism has fully formed, we get:

$w_e L^2/12 = w_p L^2/16$, thus $w_e/w_p = ¾$.

So the stiffness $K_E = 0.75*384EI/L^4 + 0.25*384EI/5L^4 =$ **$307EI/L^4$**.

7.4 SOLUTION TO EQUATION OF MOTION: FREE VIBRATION

Real systems' equation of motion: $M\ddot{x} + kx = w$

Consider free vibration, i.e. $M\ddot{x} + kx = 0$

Design of Reinforced Concrete Buildings to Resist Blast

Solution is of form $u = A\cos\omega t + B\sin\omega t$.

where A and B are determined by the boundary conditions and $\omega = \sqrt{k/M}$ and $\omega = 2\pi/T$

Proof: $u = A\cos\omega t + B\sin\omega t$

$\Rightarrow \ddot{u} = -A\omega^2\cos\omega t - B\omega^2\sin\omega t$

Substituting into equation of motion: $\Rightarrow m\ddot{u} + ku =$

$-Am\omega^2\cos\omega t - Bm\omega^2\sin\omega t + Ak\cos\omega t + Bk\sin\omega t$.

Let $\omega = \sqrt{k/m}$ ....called the "circular frequency". Thus $\omega^2 = k/m$ and $m\omega^2 = k$

$\Rightarrow m\ddot{u} + ku =$

$-Ak\cos\omega t - Bk\sin\omega t + Ak\cos\omega t + Bk\sin\omega t = 0$.

$\Rightarrow A\cos\omega t + B\sin\omega t$ is a solution.

7.5 DYNAMIC REACTIONS

Consider a simply supported beam under a UDL with uniformly distributed mass. The inertial force is I.

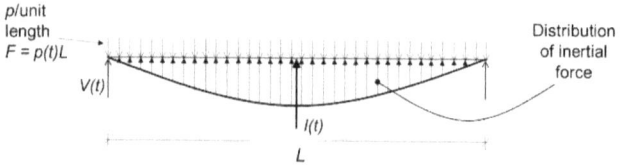

Now consider the dynamic equilibrium of the half-beam:

Design of Reinforced Concrete Buildings to Resist Blast

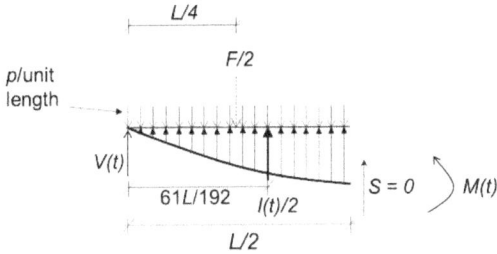

From previous work:

Deflection, $x_a = 16(L^3y - 2Ly^3 + y^4)x_{max}/5L^4$

Take moments about the resultant inertia force:

$V(61L/192) - M - \frac{1}{2} F (61L/192 - L/4) = 0$

Where M is the dynamic moment at mid-span.

Now $M = WL/8$

Where W is the resistance which varies over time.

Substituting for M,

Thus $V(t) = 0.39W + 0.11F$

Note: coefficients sum to 0.5 as equation must hold for static loading too ($W = F$). The value of F is that acting at maximum elastic displacement. For impulsive loading it is usually zero.

8

Further Examples

8.1

Summarize the reasons why an RC element can be designed to resist the enormous pressures generated by a blast.

Solution:

1. Dynamic load factor (a.k.a) amplification factor is less than 1 as t_d/T low.

2. RC is ductile when designed and detailed properly.

3. The blast duration is low.

4. Points 2 and 3 mean a plastic "collapse" mechanism can be assumed. The structure only experiences the loads consistent with this mechanism which are much lower than the loads directly induced by the blast.

8.2

Two sections have the following strain diagrams at ultimate (sagging bending). Which is more ductile?

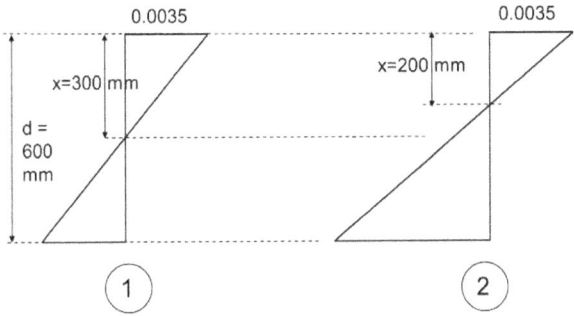

Solution:

The ultimate strain in the steel is 0.0035 for 1 and 0.007 for 2 (since by similar triangles ε_s = 0.0035*400/200). The yield strain of conventional reinforcement is $e_s = s_s/E_s$ = 0.87*500/200*10³ = 0.0022. This is less than the likely strain in the steel at ultimate in both cases (0.0035 for 1 and 0.007 for 2). Thus the steel will definitely reach yield before the concrete crushes at 0.0035. Section 2 has a smaller value of x and so is **more ductile** as a higher strain will be reached in the steel before the concrete crushes.

8.3

The following 300x300 column is subjected to blast. Calculate its moment capacity and determine its suitability for blast design. The permanent Axial Load is 1,000 kN. Static f_{ck} = 40 N/mm² and f_{yk} = 500 N/mm².

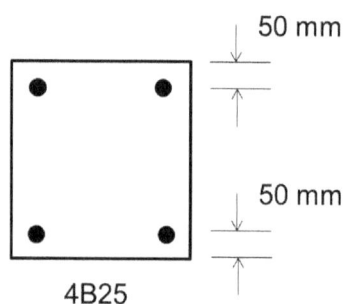

Solution

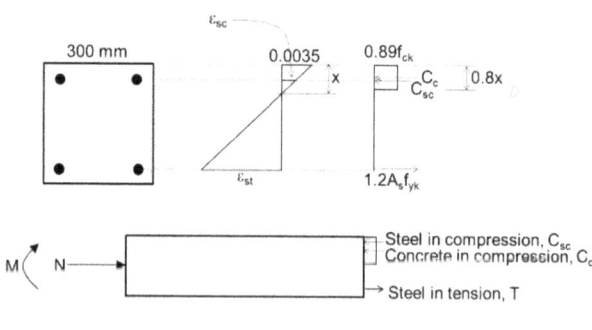

$$\sum F_x = 0 \implies N + T = C_{sc} + C_s$$

To obtain the value of x at ultimate, a trial and error procedure may be adopted:

- Guess a value of x

Design of Reinforced Concrete Buildings to Resist Blast

- Work out the ε_{sc} implied by this x
- Using E = 200,000 find C_{sc}
- Work out the total value of C (= $C_c + C_{sc}$)
- Find $T = A_s f_{ydyn}$ where $f_{ydyn} = 1.2 f_{yk}$.
- Compare with $N + T$ with C
- If within say 2-3% then stop

Using this approach the value of x is found to be about 131 mm.

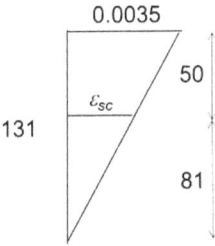

Thus the strain ε_{sc} is (81/131)0.0035 = 0.00216

Compression stress = 0.00216(200,000) = 432 N/mm²

i.e., C_{sc} = 432*491*2 = 424 kN

Now C_c = 0.89*40*0.8*131*300 = 1,119 kN

$\Rightarrow \sum C$ = 1,543 kN

Now T = 491*2*1.2*500 = 589 kN

Check equilibrium: $N + T$ = 1,589 kN ≈ $\sum C$ (within 3%)

Moment capacity, (take moments about centerline) => M_p = 210 kNm

Design of Reinforced Concrete Buildings to Resist Blast

Notice: if there is no N then and $x = 58$ mm

And $M_p \approx TZ = 589$ kN * 0.2 m = 118 kNm

Now x at ultimate is 131 mm, $d = 250$mm, thus $x/d = 131/250 = 0.52 > 0.25$. The latter is the upper limit for plastic action suggested by EC2.

Hence the section is **unsuitable** for plastic design and so blast.

The column must be enlarged. It can be shown that at least a 450x450 column is necessary for plastic action in this case.

8.4

See Bazant and Zhou (2002).

On September 11 2001 a fully laden 767-200 (mass 179,000 kg) travelling at 550 km/hr impacted the upper portion of the 110 storey tower of the World Trade Centre.

Assuming the equivalent mass M_{eq} of the upper portion of the towers is 141×10^6 kg, what is the initial velocity of the tower? Estimate the likely horizontal deflection of the tower. Model the behavior as a concentrated mass M_{eq} on a weightless cantilever. Assume elastic behavior.

Design of Reinforced Concrete Buildings to Resist Blast

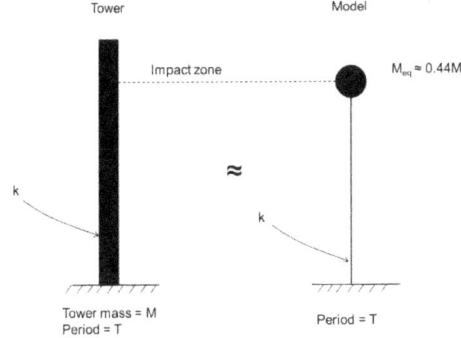

Solution

Momentum of plane before impact = momentum of building after impact, i.e. 179,000×550 = 141×10^6 v_o Thus v_o = 0.7 km/h = 0.19 m/s.

Assume the response is dominated by the first free mode of vibration, period T_1. Then the maximum deflection is $v_oT_1/2\pi$.

(Proof: The system is a free vibrating undamped SDOF subject to an initial velocity v_0. Thus equation of motion is: $m\ddot{u} + ku = 0$ which has as solution, $u = A\cos\omega t + B\sin\omega t$.

Thus $\dot{u} = -A\omega \sin\omega t + B\omega \cos\omega t$.

Now at $t = 0 \Rightarrow \dot{u} = v_0 = B\omega$ and $u = 0 = A$

$\Rightarrow u = v_0/\omega \sin\omega t$.

Thus the maximum value of $u = u_{max} = v_0/w$.

Now $T = 2\pi/\omega$, $\Rightarrow u_{max} = v_0T/2\pi$.)

Design of Reinforced Concrete Buildings to Resist Blast

Approximately, $T_1 = 14$ s

And $v_0 = 0.19$ m/s

=> $u_{max} = v_0 T_1 / 2\pi = 0.19 \times 14 / 2\pi = 0.4$ m

Now the tower was 415 m high and designed such that the wind load causes an elastic deflection of $h/500$ (= 0.83 m) at the top. Thus tower remains elastic so our method, which assumes elastic behavior, is correct.

8.5

Find the *sustained* UDL the following RC beam can support according to EC2 when the beam is designed using (a) the elastic BMD and (b) the plastic BMD. What redistribution of elastic support moments is required to get the plastic diagram?

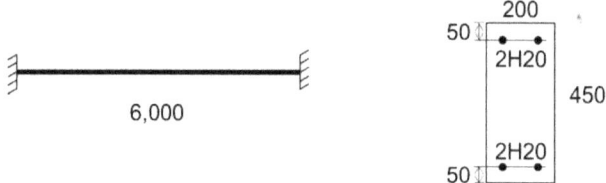

Concrete $f_{ck} = 30$ N/mm^2; Steel $f_{yk} = 500$ N/mm^2

Solution:

EC2 Rectangular stress distribution for pure bending: Sagging shown.

Design of Reinforced Concrete Buildings to Resist Blast

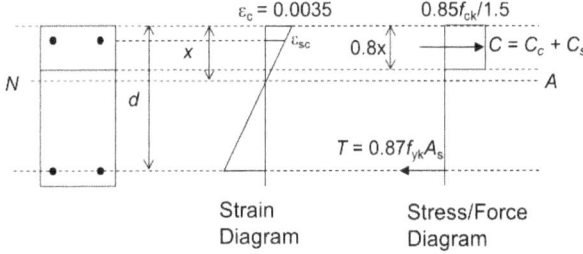

Strain Diagram   Stress/Force Diagram

$T = 0.87 \times 500 \times 2 \times 314 = 273$ kN

C = concrete force + steel force = $C_c + C_s$

(1) We guess a value of x,

(2) Then using the strain diagram we evaluate the strain in the steel, e_{sc}.

(3) Then knowing the value of E_s we can evaluate the stress in the steel, s_{sc}.

(4) We can then compare $C_c + C_s$ with T. For equilibrium they should balance. If not, select a new trial x.

x (mm)	ε_{sc}	σ_{sc}	C_s (kN)	C_c (kN)	C (kN)	$(T-C)/T$ (%)
60	0.0006	120	75.4	161.3	236.7	13.4
70	0.0010	200	125.6	188.2	313.8	-14.8
65	0.0008	161.5	101.4	174.7	276.1	-1.3

Thus it is found $x = 65$ mm is close enough.

Design of Reinforced Concrete Buildings to Resist Blast

To find Moment of Resistance of section (i.e. plastic capacity), $M_R = M_P$ take moments about any point, say top of section => M_P = $T.d - C_c(0.4x) - C_sd'$ = $273.2*10^3*400 - 174.7*10^3*(0.4*65) - 101.4*10^3*50$ = $109.3 - 4.54 - 5.07 = 99.7$ kNm.

Note: EC2 also requires us to check the suitability of this section for plastic design:

$x/d = 65/400 = 0.16 < 0.25$ => okay

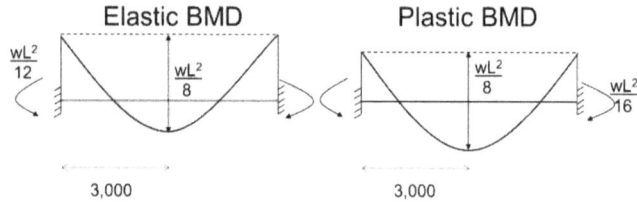

Use of the Elastic BMD gives design moments of $wL^2/12$ in the hogging region and $wL^2/24$ in the sagging,

Use of the Plastic BMD gives design moments of $wL^2/16$ in the hogging region and $wL^2/16$ in the sagging (assuming a hinges at supports and mid-span).

Elastic design:

Max value of $w = 12M_P/L^2 = 12*99.7/6^2 = 33.2$ kN/m

Plastic design:

Max value of $w = 16M_P/L^2 = 16*99.7/6^2 = 44.3$ kN/m

Design of Reinforced Concrete Buildings to Resist Blast

Redistribution of support moments required: $wL^2/12$ to $wL^2/16$, i.e. **25%**

8.6

(a) The rigid glazing system is appropriate for impulses less than about 1,400 kPa.msec.

(b) The largest impulse for which blast tests on glazing have been carried out is about 5250 kPa.msec

Given an explosive mass of 250 kg TNT what stand-off does this correspond to in each case?

Solution:

$m = 250$ kg $=> m^{1/3} = 6.3$

(a) Thus $I_r/m^{1/3} = 1400/6.3 = 222$ kPa.msec/kg$^{1/3}$

Reading from Spaghetti chart the Scaled distance corresponding to $I_r/m^{1/3} = 222$ kPa.msec/kg$^{1/3}$ is about $Z = R/m^{1/3} = 3.0$, i.e, $R = 3.0 \times 6.3 = 18.9$ m.

(b) $I_r/m^{1/3} = 5250/6.3 = 833$ kPa.msec/kg$^{1/3}$

Reading from Spaghetti chart the Scaled distance corresponding to this $I_r/m^{1/3}$ is about $Z = R/m^{1/3} = 1.0$, i.e, $R = 1.0 \times 6.3 = 6.3$ m.

Design of Reinforced Concrete Buildings to Resist Blast

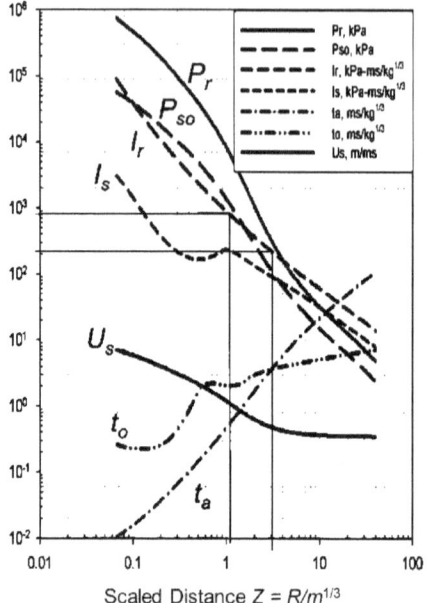

8.7

The floor at Level 2 of a substation spans 7m x 7m. It is a one-way beam and slab system. The slab is 250 mm thick and reinforced with continuous B16@200 mm each way top and bottom. Laps are 500 mm. Analysis shows a level 1 supporting column is likely to be completely destroyed by a contact blast, forcing the slab to span 14 m as a catenary.

The characteristic additional dead load is 2 kN/m² and live load is 10 kN/m². f_{ck} = 40 N/mm², f_{yk} = 500 N/mm². Determine if progressive collapse is likely.

Design of Reinforced Concrete Buildings to Resist Blast

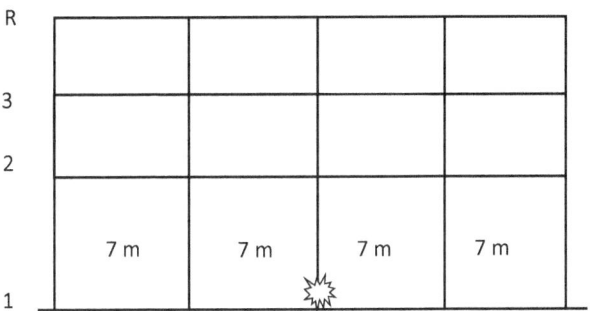

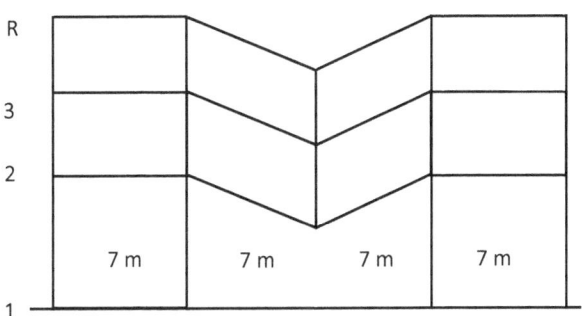

Solution:

Slab: consider 1 m width

$W_{accident}$ = DL + 0.5LL = 8.25 + 0.5x10 = 13.25 kN

Design of Reinforced Concrete Buildings to Resist Blast

We can include all the top steel as it is effectively continuous i.e. there are laps.

$T\Delta = w_{accident}L^2/8$, where $T = A_s f_{yk}$ and experiment has shown that $\Delta \approx L/10$

Required $T = wL^2/8\Delta = 13.25 \times 14^2/(8 \times 1.4) = 232$ kN/m

Allow a factor of 2 as the column is suddenly removed, thus $T = 2 \times 232 = 464$ kN.

Lap to steel required by EC2 (100% bars lapped at 1 location) = $41\phi = 656$ mm

Available $T = A_s f_{yk} = 2 \times 201 \times 5 \times 500 \times (500/656) = 766$ kN

> 464 kN.

Thus it is likely the floor at this level can survive. Remaining floors as well as the horizontal load should be checked to ensure progressive collapse is unlikely.

8.8
(Hambly, 1994)

When analysing a building's progressive collapse resistance, the vierendeel mechanism's contribution must often be evaluated. The following example concerns this mechanism. A steel vierendeel truss is shown below. Use approximate analysis to estimate the maximum bending moment in the members. Compare the maximum deflection of a beam having a continuous

Design of Reinforced Concrete Buildings to Resist Blast

web with the vierendeel deflection, which is found by computer analysis to be a maximum of is 43 mm (at mid-span).

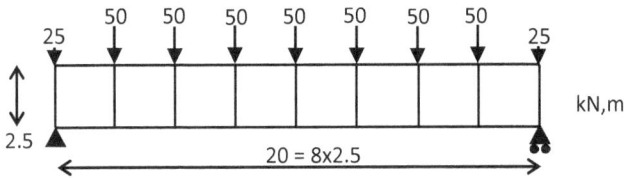

Horizontals: $A = 0.15$ m^2; $I = 0.000177$ m^4

Verticals: $A = 0.12$ m^2; $I = 0.000016$ m^4

All members $E = 210$ GPa

Solution:

Assume a point of contraflexure at the mid-point of each member, i.e. hinges placed as follows:

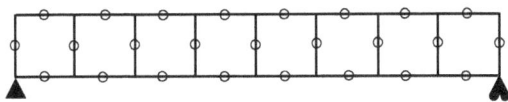

The structure is now statically determinate.

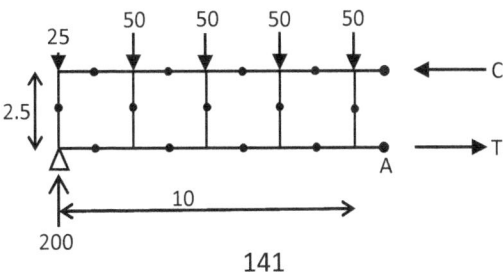

Design of Reinforced Concrete Buildings to Resist Blast

$M_A = 0 \Rightarrow C = (175 \times 11.25 - 50 \times (8.75+6.25+3.75))/2.5$

$= 412.5$ kN

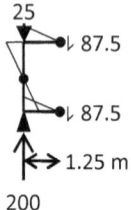

$M_{max} = 87.5 \times 1.25 = 109.4$ kNm

Computer solution: $C = 360$ kN; $M_{max} = 120$ kNm

Deflection of "equivalent" beam:

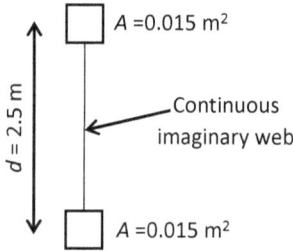

UDL = $8 \times 50/20 = w = 20$ kN/m

Midspan deflection, $\delta = 5wL^4/384EI_{beam}$ where = $I_{beam} = Ad^2/2 = 0.015 \times 2.5^2/2 = 0.047$ m^4

Thus $\delta = 5 \times 20 \times 20^4/(354 \times 210 \times 10^6 \times 0.047) = 0.004$ m.

Computer solution to vierendeel: $\delta = 0.043$ m, i.e. nearly 10 times! This indicates how flexible this load resisting mechanism is.

Design of Reinforced Concrete Buildings to Resist Blast

8.9

Derive EC2 safety factors for concrete and steel for normal loads and blast.

Note EC2 treats the blast load as an "accident".

The factors further allow for the dynamic strength of the materials.

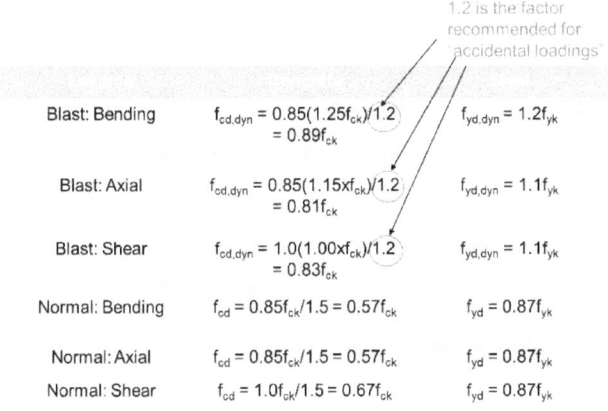

Blast: Bending	$f_{cd,dyn} = 0.85(1.25f_{ck})/1.2$ = $0.89f_{ck}$	$f_{yd,dyn} = 1.2f_{yk}$
Blast: Axial	$f_{cd,dyn} = 0.85(1.15xf_{ck})/1.2$ = $0.81f_{ck}$	$f_{yd,dyn} = 1.1f_{yk}$
Blast: Shear	$f_{cd,dyn} = 1.0(1.00xf_{ck})/1.2$ = $0.83f_{ck}$	$f_{yd,dyn} = 1.1f_{yk}$
Normal: Bending	$f_{cd} = 0.85f_{ck}/1.5 = 0.57f_{ck}$	$f_{yd} = 0.87f_{yk}$
Normal: Axial	$f_{cd} = 0.85f_{ck}/1.5 = 0.57f_{ck}$	$f_{yd} = 0.87f_{yk}$
Normal: Shear	$f_{cd} = 1.0f_{ck}/1.5 = 0.67f_{ck}$	$f_{yd} = 0.87f_{yk}$

1.2 is the factor recommended for accidental loadings

8.10

Use the Hader diagram to assess the damage to column G24 of the Alfred Murrah building, Oklohoma.

Stand-off = 4.75 m

Explosive mass = 1,800 kg TNT

Column dimension = 0.915x0.51 m

Solution:

$t/m^{1/3} = 0.51/1800^{1/3} = 0.04$

$r/m^{1/3} = 4.75/1800^{1/3} = 0.39$

The diagram indicates "perforation".

In fact, the column was actually completely destroyed by brisance. (See for example MacAlevey 2010).

8.11

A wall panel spans 6 m vertically. It is subject to a blast of 250 kg TNT at a 28 m stand-off giving a reflected pressure-time diagram as follows:

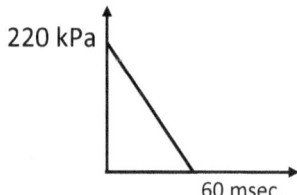

The cladding panel is overall 300 mm thick and can be assumed to be built-in at both supports and its dynamic moment capacity taken as 414 kN per m width. Concrete

Design of Reinforced Concrete Buildings to Resist Blast

static strength, f_{ck} = 35 N/mm² (E = 34 GPa). Steel is B500B (i.e. f_{yk} = 500 N/mm²) and ρ = ρ' = 0.012 and d = 260 mm.

Check the flexural design assuming category 1 is required. (Shear design follows examples given already).

Solution:

The first thing to notice is that the duration of the load (60 ms) is much longer than anything we have considered up to now. In fact the loading is still dynamic but it's NOT impulsive, as we'll show later.

Initially we follow the same procedure, finding w, X_m and X_E as before, but we don't need to satisfy the impulse equation.

Given M_p = 414 kNm

Fixed-ended beam, thus $w = 8(M_{HOG}+M_{SAG})/L^2$

= 8(2*414)/6² = 184 kN/m per m width.

Protection category 1 (i.e. 2°) => X_m = 3tan2° = 0.105 m.

We find X_E by first finding I using the graph previously used.

α = 200/34 = 5.88 and ρ = 0.021.

Thus I = 0.044bd^3 = 0.044*1*0.26³ = 0.00077 m⁴.

$K_E = 307EI/L^4$ = 307*34*10⁹*0.00077/6⁴ = 6.23*10⁶ N/m

Design of Reinforced Concrete Buildings to Resist Blast

$X_E = w/K_E$ so $X_E = 184*10^3/6.23*10^6 = 0.03$ m

Now for a SDOF system $T = 2\pi/\omega$ and $\omega = \sqrt{(k/M)}$

Thus $T = 2\pi \sqrt{(M/k)}$

However, in our case we have an equivalent SDOF system to the real one. Thus we use $T = 2\pi \sqrt{(K_{LM}M/LK_E)}$

$K_{LM} = (0.77 + 0.66)/2 = 0.72$

And $M = \gamma_{conc}tL = 2,500*0.3*6*1 = 4,500$ kg

Thus $T = 2\pi\sqrt{(0.72*4,500/(6*6.23*10^6)} = 59$ ms

We now proceed to read from the following 2 charts.

The first indicates all possible solutions given values of w/p_r (= 184/220 = 0.84) and t_d/T (= 60/59 = 1.01).

Hence min $X_m/X_E = 3.2$ and $X_m = 3.2*0.03 = 0.096$ m

< 0.105 so flexural capacity okay.

Design of Reinforced Concrete Buildings to Resist Blast

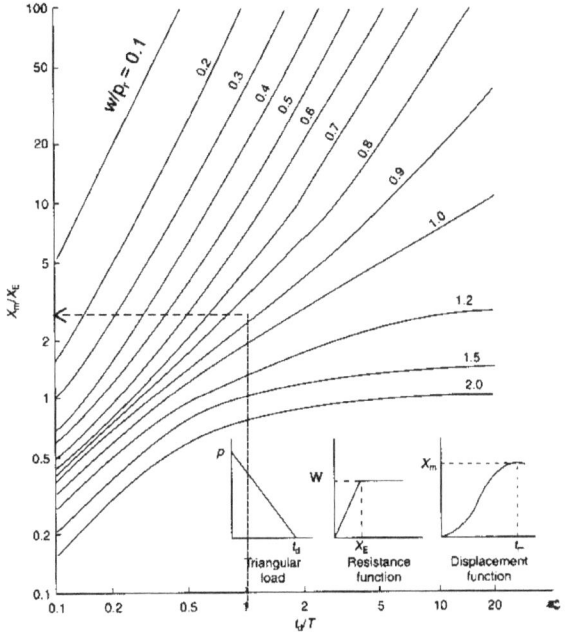

From the second chart we note that t_m/t_d = 0.7 which is < 3 (our limit for impulsive behaviour). Thus the load regime is "dynamic".

Design of Reinforced Concrete Buildings to Resist Blast

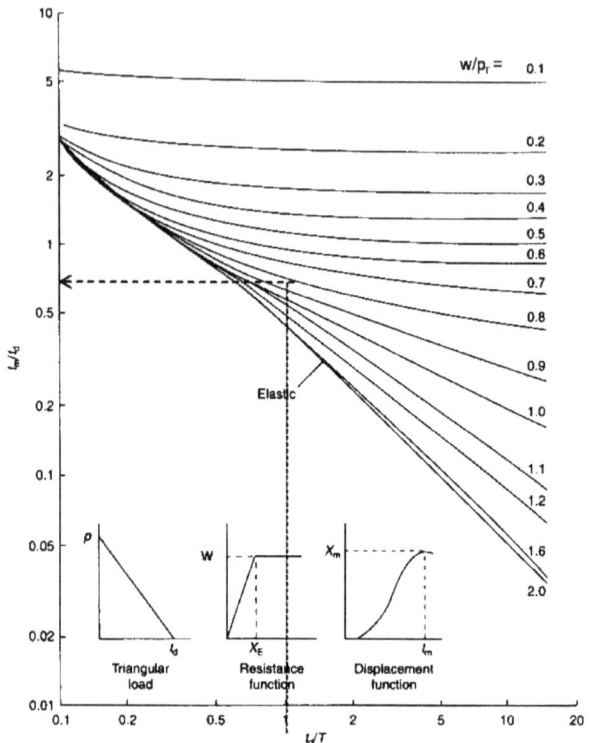

8.12

Rework the previous example assuming impulsive behavior. Comment on whether the solution obtained is conservative or unconservative.

$i^2 A^2 / 2K_{LM} M = W(X_M - X_E/2)$

$i = 220 \times 60/2 = 6,600$ kPa.msec; $A = 6$ m²; $K_{LM} = 0.72$; $M = 2500 \times 0.3 \times 6 = 4,500$ kg.

$W = 184 \times 6 = 1,104$ kN; $X_M = 0.105$ m; $X_E = 0.03$ m.

LHS (i.e. energy input) = 242,000 Nm.

Design of Reinforced Concrete Buildings to Resist Blast

RHS (i.e. energy that can be absorbed) = 99,360 Nm.

Thus our conclusion would be that the solution does not work and "W" needs to increase substantially in this case.

Thus the approach of assuming impulsive behavior is conservative (of course wasteful). The attractiveness of this assumption is that hand-based calculations are then possible.

Design of Reinforced Concrete Buildings to Resist Blast

References:

ACI/SEI 59-11, *Blast Protection of Structures*, USA, 2011.

American Concrete Institute, ACI-318-11: Building Code Requirements for Structural Concrete and Commentary, Michigan, USA, 2011.

Bažant, Z. P. and Zhou, Y, *Why Did the World Trade Center Collapse?—Simple Analysis,* ASCE Journal of Engineering Mechanics, January 2002. *(Available free on-line.)*

Biggs, J. M., *Introduction to Structural Dynamics,* McGraw-Hill, New York, 1964.

British Standards Institution, BS EN 1992-1-1:2004. Eurocode 2: Design of Concrete Structures. General Rules for Buildings, London. 2004.

Cormie, D., Mays, G., and Smith, P., *Blast Effects on Buildings*, 2nd edition, Thomas Telford, London, 2009.

Department of Defense Manual: *"Antiterrorism Standards for Buildings (UFC4-010-01)"*, 2007".

Department of Defense, *Structures to Resist the Effects of Accidental Explosions* (United Facilities Criteria, UFC 3-340-02), USA, 2008. Download free.

Dusenberry, D., *Handbook for Blast Resistant Design of Buildings*, J. Wiley, USA, 2010.

GSA TS01-2003, US General Services Administration *Standard Test Method for Glazing and Window Systems Subject to Dynamic Overpressure Loadings.*

Hader, H., *Effects of Bare and Cased Explosive Charges on Reinforced Concrete Walls,* Ernst Basler & Partners Consulting Engineers, Switzerland, 1983.

Hambly, Edmund C., *Structural Analysis by Example*, Archimedes, 1994.

Humar, J. L., *Dynamics of structures*, 2^{ed}, Taylor & Francis, London, 2002.

Kennedy, C, Goodchild, C, *Practical yield line design*, Reinforced Concrete Council. BCA, 2003. Available free at www.rcc-info.org.uk.

Krauthammer, T., *Modern Protective Structures*, CRC Press, USA, 2008.

MacAlevey, N., *Structural Engineering Failures: lessons for design*, CreateSpace, 2010.

MHA, *Guidelines for Enhancing Building Security in Singapore* (GEBSS), 2010. Download free.

Paulay, T., and Priestly, M.J.N., *Seismic Design of Reinforced Concrete and Masonry Buildings*, John Wiley & Sons, inc, USA, 1992.

Smith, P.D. and Heatherington, J.G., *Blast and Ballistic Loading of Structures*, Butterworth Heinemann, UK, 1994.

Tian, Y., Su, Y., *Dynamic Response of Reinforced Concrete Beams Following Instantaneous Removal of a Bearing Column*, International Journal of Concrete and Materials, Vol. 5, No. 1, pp. 19-28, June 2011.

Design of Reinforced Concrete Buildings to Resist Blast

US Department of Army, *Explosives and Demolition*, Field Manual 5-250, 1992. Download free.

If you have any comments or queries related to the book please feel free to contact me at: niallmacalevey@gmail.com

Design of Reinforced Concrete Buildings to Resist Blast

Notes

Design of Reinforced Concrete Buildings to Resist Blast

Notes

www.ingramcontent.com/pod-product-compliance
Lightning Source LLC
Chambersburg PA
CBHW071800200526
45167CB00017B/535